Army Aviation Special and Incentive Pay Policies to Promote Performance, Manage Talent, and Sustain Retention

AVERY CALKINS, MICHAEL G. MATTOCK, BETH J. ASCH,
RYAN A. SCHWANKHART, TARA L. TERRY

Prepared for the United States Army
Approved for public release; distribution unlimited

For more information on this publication, visit **www.rand.org/t/RRA2234-1**.

Published by the RAND Corporation, Santa Monica, Calif.

Library of Congress Cataloging-in-Publication Data is available for this publication.
ISBN: 978-1-9774-1119-8

Cover: Matt Hecht/U.S. Department of Defense.

About This Report

This report documents research and analysis conducted as part of a project entitled *Optimizing Army Aviation Retention Through Efficiency and Effectiveness of Aviation Special and Incentive Pay*, sponsored by the U.S. Army. The purpose of the project was to help the Army determine how to modernize special and incentive pays to better reward qualified Army aviators while cost-effectively achieving retention objectives.

This research was conducted within RAND Arroyo Center's Personnel, Training, and Health Program. RAND Arroyo Center, part of the RAND Corporation, is a federally funded research and development center (FFRDC) sponsored by the United States Army.

RAND operates under a "Federal-Wide Assurance" (FWA00003425) and complies with the *Code of Federal Regulations for the Protection of Human Subjects Under United States Law* (45 CFR 46), also known as "the Common Rule," as well as with the implementation guidance set forth in DoD Instruction 3216.02. As applicable, this compliance includes reviews and approvals by RAND's Institutional Review Board (the Human Subjects Protection Committee) and by the U.S. Army. The views of sources utilized in this study are solely their own and do not represent the official policy or position of DoD or the U.S. Government.

Acknowledgments

We are grateful to Mr. Vo Van in the U.S. Army Training and Doctrine Command who sponsored our project. We would very much like to thank CW5 Steve Donahue in the U.S. Army Aviation Center of Excellence who served as project monitor, provided background material, and served as a tremendous resource for this project. We are also very grateful to CW4 Damon Hutton in the Army's Human Resources Command who answered our questions concerning identifying pilot career milestones in the Army personnel data. We also gratefully acknowledge the help of Tony Lawrence and Daniel Schwam who both provided research programming support, and we wish to thank our RAND colleagues Heather Krull and Katharina Best, the director and associate director, respectively, of the Personnel, Training, and Health Program within the Arroyo Center. Finally, our research benefited from the input of two reviewers, Patricia Tong of RAND and Curtis Simon of Clemson University.

Summary

Commissioned officers and warrant officers in U.S. Army Aviation are eligible for aviation incentive pay (AvIP), which is a monthly special pay meant to maintain an inventory of trained aviators. AvIP, along with the aviation bonus (AvB), are the two main special and incentive (S&I) pays for officers who are aviators (although only warrant officers are eligible for AvB in the Army). AvB is a financial incentive offered to eligible aviators to remain in service for an additional service obligation. As part of its talent management strategy, the Army is looking to modernize its S&I pays in general to increase their efficiency in improving retention and their incentives for greater performance. To modernize these S&I pays for aviators, the Army is considering proposals that would make these pays contingent on achieving specific career milestones. Such a policy change could increase incentives for the development of human capital and improve retention among those who achieve those milestones. Thus, the aim of the proposal would be not only to sustain retention but also to target compensation to individual qualifications and talent.

In addition, the Army is considering the impact of changes in the civilian economy and in the military retirement system on the effectiveness of Army aviator S&I pays. Because of high demand for pilots in the civilian economy, as we discuss in this report, retention of aviators is an ongoing concern for the Army, leading to questions of how S&I pays for Army aviators might be reformed to sustain aviator retention. The nominal value of AvIP was not updated between 1998 and 2020, degrading its value in real dollars. Although AvIP was updated in 2020, the same issue of a substantial drop in the real value of AvIP could recur if the Army fails to account for the historic levels of inflation in 2022. A substantial drop in the real value of AvIP could weaken the incentive to stay in the Army as an aviator. Improvements in civilian pay and lower unemployment in the civilian economy increase the attractiveness of external opportunities, raising questions about the extent to which increases in AvB could offset any negative retention effects. Finally, the Army has expressed concern that the new blended retirement system (BRS), which covers service members who entered the Army as of January 1, 2018, and any current service member covered by the legacy system who chose to opt into the new system, could hurt aviator retention.

The Army requested that RAND Arroyo Center provide analyses to improve the setting of these S&I pays for aviators, including conducting an analysis of whether AvIP could be set not just to sustain retention but to be based on achievement of key career milestones. Development of models to assess alternative proposals to set S&I pay based on milestones could also facilitate analyses of the retention effects on aviators of other potential compensation policies, including increasing AvIP to keep up with inflation, varying the AvB in response to changes in the civilian labor market pay, and the implementation of the new BRS. This report summarizes that analysis and includes an empirical assessment of these alternative policies, including reform proposals that would set AvIP based on career milestones, and examines how these proposals would affect Army aviator retention. The policies we considered include

- setting AvIP based on career milestones rather than years of aviation service (YAS), the current method for setting AvIP
- keeping the real value of AvIP constant and showing the effect of the fall in the real value of AvIP from 1998 to 2020
- changing the size of AvB under alternative scenarios of changes in civilian earnings and unemployment
- switching the initial active-duty service obligation (ADSO) from six years to ten years to align with the Army's switch to longer aviator ADSOs in fiscal year (FY) 2021
- changing from the legacy retirement system to the BRS.

Approach

The modeling effort described in this report builds on previous research on the effects of compensation on retention in the military that use the dynamic retention model (DRM), focusing on studies that considered contracts that impose a multi-year service obligation in exchange for a bonus (Asch et al., 2019; Hosek et al., 2017; Mattock et al., 2016; Mattock and Arkes, 2007). We updated the DRM to estimate the parameters underlying the retention decisions of Army warrant officer and commissioned officer pilots. Because many, though not all, warrant officer aviators have prior enlisted service, we updated the model to include prior enlisted service for warrant officers. We also incorporated aviation career milestones using Army data on the achievement of such milestones, and we considered alternative specifications of the potential external earnings opportunities available to aviators should they leave the Army. To estimate the model, we used the Army's Total Army Personnel Database for officers, which allowed us to track the individual career histories of officer aviators who entered the Army between 2002 and 2010 through 2021. We also made use of information on military pay and the history of S&I pays offered to Army aviators. To capture civilian opportunities, we analyzed profiles of earning by age derived from the American Community Survey for pilot and non-pilot veterans with some college, as well as pilot and non-pilot veterans with four years of college. As part of our assessment of the civilian opportunities available to Army pilots, we also considered available evidence on the demand for pilots by airlines, including pilots of both fixed-wing and rotary-wing aircraft.

We estimated a separate model of warrant officer and commissioned officer aviators and found our models fit the data well. We then developed a capability that enabled us to simulate the retention effects of alternative S&I pays for Army aviators and other policy changes.

As part of our assessment of the retention effect of setting AvIP based on career milestones, we analyzed the extent to which such a policy would increase the retention of higher-ability pilots, especially at more senior career milestones. This assessment built on two previous studies that analyzed the effects of changes in military compensation on the retention of higher-ability personnel (Asch et al., 2021; Asch, Mattock, and Tong, 2020).

We note that our work has several limitations. First, our simulations of the retention effect of AvIP based on career milestones depended on our assumptions regarding the timing and probability of achieving career milestones. While we based these assumptions on observed values in the Army data, we also performed sensitivity checks and found that our estimates were not sensitive to our assumptions. Second, we did not observe or simulate the incentive for aviators to acquire additional human capital when AvIP depends on achieving aviation milestones as noted above. Third, our assessment of the retention effects of different policies did not consider how these policies would affect the *taste distribution* (a function that gives the probability of occurrence of a particular level of individual preference [expressed in dollars] for active service) of the entry cohort and, thereby, affect subsequent retention.

Key Results

Setting AvIP Based on Career Milestones Will Require Higher Than Current AvIP Levels to Sustain Current Retention

The advantage of setting AvIP based on milestone achievement is that it increases the incentive of aviators to achieve the milestones. It also increases retention incentives as aviators look to the future and assess their likelihood of achieving those milestones and the higher AvIP associated with each milestone. A disadvantage is that not all Army aviators who remain in the Army to each YAS achieve the key career milestones, and so, not all aviators would earn higher AvIP if AvIP depends on career milestones achievement. Under current policy, all aviators who reach a given YAS earn higher AvIP. The new policy could reduce the retention effect of AvIP because the likelihood of being paid a higher AvIP is lower. Table S.1 shows the timing and probability of milestone achievement for commissioned officer and warrant officer aviators, where the probability is conditional on staying in the Army long enough to achieve the milestone.

The Army provided us with a set of prototype dollar values as a starting point for examining the retention effects of setting AvIP based solely on career milestone achievement; these values were similar to the AvIP values that the Army currently uses but are based on YAS. For example, under current policy, monthly AvIP for aviators with 2 YAS equals $200. Under the prototype, personnel who achieve pilot status, typically at around 1.5 YAS (rounded up to 2 in our model), would receive $200 monthly.

Our analysis confirms our intuition regarding the retention effects of basing AvIP on milestones. We found that setting AvIP values based on career milestones equal to the current policy, as under the Army's prototype, would reduce retention (measured as the change in the size of the aviator inventory) by approximately 1.0 percent for commissioned officers and 1.9 percent for warrant officers. This policy would also reduce AvIP cost to the Army, primarily because retention is lower and not all aviators under the new policy at a given YAS would receive higher AvIP. We then identified values of milestone-based AvIP that would maintain retention (that is, values that lead to the same overall end strength and similar retention at each YAS). We found that the Army would need to offer substantially higher values of AvIP than the prototype if it wishes to maintain retention, though we note that those values may not be unique. The higher AvIP values would also increase total AvIP cost relative to current policy.

The retention-maintaining AvIP would not be the same for commissioned officers and warrant officers. Under current policy, AvIP is the same dollar amount for both groups. The values that sustain retention for commissioned officers would be higher than for warrant officers, owing to the different retention behavior of each group and differences in their responsiveness of changes in compensation. Should the Army decide to set the same AvIP levels for both groups, based on milestones, our analysis indicates that retention would change for one of the two groups. If the values were set at the retention-maintaining level for commissioned officers, it could lead to higher costs and economic rents paid to warrant officers, meaning they would receive higher pay than required to sustain their retention relative to the baseline.

We also found that the retention-maintaining values of an AvIP based on milestones would not need to increase as much, and the increase in cost would be lower if Army aviators had a higher likelihood of achiev-

TABLE S.1

Timing and Probability of Milestone Achievement

	Commissioned Officer Aviators		Warrant Officer Aviators	
Milestone	Milestone Timing[a]	Probability of Achievement[b]	Milestone Timing[a]	Probability of Achievement[b]
Pilot status	2 YAS	0.97	2 YAS	0.98
Pilot-in-command	4 YAS	0.65	4 YAS	0.91
Captain's Career Course (commissioned officers only)	6 YAS	0.98	NA	NA
Track (warrant officers only)	NA	NA	6 YAS	0.68
Senior aviator	8 YAS	0.56	10 YAS	0.65
Master aviator	16 YAS	0.24	16 YAS	0.54

[a] Milestone timing is rounded to the nearest YAS because these timings are used as inputs to the DRM.

[b] Probability of achievement is the probability that an individual achieved the milestone at or before the milestone timing YAS plus two years, conditional on having stayed in the Army until the year of the milestone plus two years and having achieved all previous milestones (except pilot status, whose probability was estimated at 2 YAS).

ing career milestones than what we observed in the data and used in our modeling effort. Thus, to the extent that basing AvIP on milestones increases the likelihood that a pilot achieves the milestone, the increase in cost and the degree to which AvIP must be higher are reduced.

Setting AvIP Based on Career Milestones Can Increase the Incentives for Higher-Ability Pilots to Remain in Service

Another potential advantage of basing AvIP on milestone achievement is its effect on the retention of higher-ability personnel. To the extent that higher-ability aviators achieve these milestones faster, basing AvIP on milestones also increases the incentive for higher-ability aviators to stay in service. We considered two additional proposals that base AvIP on career milestones. Both have the feature that higher-ability personnel who achieve the milestones sooner receive more AvIP than lower-ability personnel who achieve the milestones later. Under the first variant, the value of AvIP increases in a way that is proportional to the retention-maintaining AvIP schedule. Under the second variant, AvIP values double at each successive milestone. Under these proposals that increase the reward for achieving milestones, especially for those who achieve milestones faster, we found that the retention of higher-ability personnel increases. We also found that *ability sorting* (higher-ability individuals sorting to higher milestones) improves—that is, relatively more high-ability personnel are retained at later milestones under these proposals than under the baseline current system. Not only were the individuals retained of higher ability on average at each milestone, but with each additional milestone, the improvement in mean ability over the baseline also increased.

Retention Increases When the Initial Service Obligation Is Increased

As expected, we found that increasing the initial ADSO to ten years substantially increases retention over the entire aviator career, not just during the first ten years. For commissioned officers, the force size in the steady state is predicted to rise by 25.4 percent while the increase for warrant officers is predicted to be 14.9 percent. The overall increase in force size is likely because of members being four years closer to vesting in the defined-benefit retirement annuity, while the difference in the increase between commissioned officers and warrant officers would be due, in part, to differences in discounting (time preference)—i.e., commissioned officers value the retirement annuity more. As noted above, these estimates do not consider how such a policy would affect the taste distribution of entrants, the number of accessions, or the quality of entrants. Increasing the initial ADSO has the effect of lowering the value of a career in Army Aviation from the point of view of a potential entrant, because service members no longer have the option to leave after their sixth through ninth years. Other things equal, this change will tend to result in an entering population with a higher taste for Army Aviation.

Increases in AvB Can Offset the Adverse Effects of Civilian Demand

We also examined retention responses to higher AvB levels. We first assumed a base case where no AvB is offered. We found that offering $15,000 in AvB to tracked warrant officers (those who are safety, maintenance, mission survivability, or instructor pilot officers) only (which is roughly half the amount of AvB offered in FY 2020 and FY 2022) can maintain Army Aviation retention even in the event of significant improvements in external opportunities, such as a drop to 0-percent unemployment or a 10-percent increase in civilian pay relative to military pay. We found similar results when our base case is adjusted so that a $30,000 AvB is offered—that is, offering an additional $15,000 ($45,000 total) could address sizable shocks to external opportunities.

By Increasing Continuation Pay Above the Minimum Level, the Army Can Offset Adverse Retention Effects of the Blended Retirement System

We used the simulation capability in our aviator models to simulate how BRS would affect aviator retention. Continuation pay is one of the three major elements of the BRS and is paid to service members between completion of 8 years of service (YOS) and completion of 12 YOS who agree to at least three years of obligated service. The other two elements of BRS are a defined-contribution plan (matching up to 5 percent of service member contributions to a thrift savings plan) and a defined-benefit plan vesting at 20 YOS (offering an annuity of the average highest three years of basic pay). The legacy retirement system offered solely a defined benefit vesting at 20 YOS with the average highest three years of basic pay. We compared retention assuming the minimum value of continuation pay (as set by Congress) versus a continuation pay level that would sustain retention relative to the baseline. We found that under the BRS, the minimum continuation pay multiplier of 2.5 is not sufficient to sustain retention for either commissioned officer or warrant officer aviators relative to the legacy retirement system. However, we found that by raising the multiplier to 9.0, the Army could sustain aviator retention for commissioned officers, and raising the multiplier to 4.50 (if AvB is not offered) or 3.25 (if AvB is offered) will restore aviator retention for warrant officers.

Conclusions

Our analysis indicates that setting AvIP based on career milestones could sustain Army aviator retention but would increase AvIP costs and overall personnel costs to the Army (as measured by the cost of regular military compensation, military retirement, and aviation S&I pays for personnel who remain in the Army). If the Army wants to maintain current retention levels, AvIP amounts need to be higher than current levels to offset the fact that not all aviators will achieve the key career milestones and so would not be eligible for the higher AvIP amounts associated with reaching those milestones. We found that an increase in the likelihood of achieving milestones would mean that costs would increase by less. Potentially offsetting the higher costs is the greater retention of aviators with higher abilities. Better ability sorting could offset the higher costs, but these improvements only occurred when AvIP values sufficiently reward the achievement of milestones by higher-ability personnel.

Contents

Figures and Tables

Figures

Tables

CHAPTER 1

Introduction

As part of its talent management strategy, the U.S. Army is considering how special and incentive (S&I) pays might be modified to improve the performance of personnel and increase overall efficiency. According to the Defense Advisory Committee on Military Compensation (2006), S&I pays should promote performance and talent management, as well as sustain retention and recognize the unusual risks and rigors that service members face in some assignments and occupations. Aviation S&I pays help the Army sustain an inventory of pilots by recognizing persistently higher civilian pay for pilots than for non-pilots, dangerous working conditions, temporary fluctuations in external supply and demand conditions, and high training costs. These pays have become particularly important in recent years because of the growth in hiring among major airlines. In accordance with federal law and U.S. Department of Defense (DoD) policy, the two S&I pays that the Army may offer eligible officers are aviation incentive pay (AvIP) and aviation bonuses (AvBs) to continue aviation service (Department of Defense Instruction [DoDI] 7730.67, 2016). As we explain in more detail in Chapter 2, AvIP is a monthly payment to all eligible officers, the amount of which depends on the officer's year(s) of aviation service (YAS), and does not depend on reaching any specific career milestone. The current values of AvIP were established in 2020 for Army officers, but, as we show in Chapter 2, the real value of AvIP has not kept up with inflation since 1998 when AvIP values were set. In contrast, AvB is offered selectively to qualified pilots in critical aviation specialties when the Army faces a shortage of pilots. Unlike AvIP, receipt of AvB requires that officers incur an additional obligation to serve longer in the Army.

To modernize these S&I pays for aviators as part of its talent management effort, the Army is considering proposals that would make these pays contingent on achieving specific career milestones, such as achieving pilot-in-command status or enrolling for professional military education (PME). Making pays contingent on these milestones could enhance incentives for officers to improve their human capital, which is valuable to the Army. The aim of the reform would be to sustain retention, particularly of highly-talented and highly-qualified officers. For any reform proposal to move forward with the other services, the Office of the Secretary of Defense (OSD), and Congress, analysis is needed to determine how this reform would affect active-component (AC) aviator retention of both warrant officers and commissioned officers; costs; and performance measured in terms of career development, the acquisition of PME, and the retention of better-qualified aviators.

The Army requested that RAND Arroyo Center provide analyses to improve the setting of these S&I pays for aviators, including analyses of the effects of setting S&I pays based on career milestones. The development of models to assess alternative proposals to set S&I pays based on milestones could also facilitate analyses of the retention effects on aviators of other potential compensation policies, such as increasing AvIP to keep up with inflation, varying AvB in response to changes in the civilian labor market pay, and implementing the new blended retirement system (BRS), which covers officers who entered the Army as of January 1, 2018 and any current service member covered by the legacy system who chose to opt into the new system. This report summarizes these analyses and includes an empirical assessment of these alternative policies, including reform proposals that would set AvIP based on career milestones and affect Army aviator retention.

Approach

Ideally, to evaluate alternative approaches to setting aviator S&I pays and other compensation policies on retention and cost, we would use a randomized control trial methodology that randomly assigned Army aviators to control and test cells, and outcomes would be assessed before and after the assignment. Because such evaluations are costly and time consuming, we developed a modeling capability that is well suited to provide quantitative estimates of the effects of policies that do not currently exist or for which insufficient time has passed to permit an analysis of the long-term effects of the policy. This modeling capability is known as the dynamic retention model or DRM (Mattock et al., 2016; Asch et al., 2008; Gotz and McCall, 1984). The RAND Corporation has developed and used the DRM for many studies concerned with how compensation reform affects retention, including an analysis of offering a Critical Skills Retention Bonus to special operations officers in the Army (Asch et al., 2019), an analysis of raising the cap on aviation bonuses for military pilots (Mattock et al., 2019), and more recently, an analysis of reforming Army selective reenlistment bonuses for enlisted personnel to increase performance incentives (Asch et al., 2021).

For our study of Army aviator S&I pay, we estimated DRM models for commissioned officer aviators and for warrant officer aviators. We created longitudinal data on individual officers using personnel records from the Total Army Personnel Database (TAPDB) for officers. These longitudinal data track soldiers' careers from entry through 2021, starting in the year 2002. Given the estimated parameters of the DRM models for commissioned officers and warrant officers, we then developed computer code to simulate how the current aviator S&I pay approach and reform alternatives would affect retention. We also assessed how maintaining the real value of AvIP would affect Army aviator retention, the effects of varying AvB on retention, and the predicted effects of the BRS under different levels of continuation pay. Continuation pay is one of the three major elements of the BRS and is paid to service members at 12 years of service (YOS) who agree to at least three more years of obligated service.

Organization of This Report

Chapter 2 provides an overview of the careers of Army aviators, key milestones within those careers, and more detailed information on AvIP and AvB. Chapter 3 discusses the civilian opportunities available to Army aviators either as fixed-wing or rotary-wing pilots or in non-pilot occupations. Chapter 4 describes the DRM, provides parameter estimates, and shows how well the model fits historical data. Chapter 5 applies the estimated model to alternative AvIP schedules and shows how well these schedules perform with respect to cost and retaining high-ability personnel. Chapter 6 applies the estimated model to understand the effects of offering AvB on retention, especially in the presence of economic changes. Chapter 7 applies the estimated model to understand the steady-state implications of two recent policy changes: the move to a ten-year ADSO and the implementation of the BRS. In Chapter 8, we offer our concluding remarks, and the appendixes provide further detail of our methods and the data used.

CHAPTER 2

Overview of Army Aviation Careers and Special and Incentive Pays

To understand how basing S&I pays on career milestone achievements will affect retention, we required information about aviator career progression (especially the key milestones) and S&I pays, including the extent to which Army aviators have historically received those pays and achieved those milestones. This chapter begins with an overview of the career progression of Army aviators and then discusses the setting of aviation S&I pays in the Army. Tabulations related to the achievement of these milestones are shown in Appendix A, while receipt of these pays is shown in Appendix B. We also briefly discuss these tabulations in this chapter. Appendix A provides further details pertaining to active aviator careers. This chapter and the referenced appendixes include many terms specific to Army Aviation. A list of abbreviations and a glossary can be found at the end of the report.

Aviation Career Overview

Description of Pilots' Careers

Both commissioned officers and warrant officers can serve as Army aviators. Commissioned officers are commissioned via the U.S. Military Academy, the Reserve Officer Training Corps, attendance at Officer Candidate School after college graduation, or direct commission after earning a professional degree. Warrant officers are sourced from the U.S. Army Warrant Officer Selection Program, Civilian Warrant Officer Flight Training, and in-service non-aviation commissioned officers. Warrant officers in Army Aviation differ from warrant officers in other parts of the Army in that some do not have prior enlisted service, and among those who have previously served in the enlisted force, there is more variation in the number of years of enlisted service than found in other parts of the Army. Entry to the Army Aviation Branch incurs an active-duty service obligation (ADSO), which was six years prior to fiscal year (FY) 2021 and ten years thereafter (Rempfer, 2020; U.S. Army, undated).

Our understanding of the milestones of aviation careers and the ideal YAS of milestone achievement is based on discussions with our sponsor's office, the U.S. Army Aviation Center of Excellence within U.S. Army Training and Doctrine Command. Our sponsor also provided a graphical description of the ideal timing of different aviation career milestones as of 2022, reproduced in Figures 2.1 (commissioned officers) and 2.2 (warrant officers).[1] Following entry to Army Aviation, both commissioned and warrant officers enter initial entry rotary wing (IERW) training (also referred to as flight school), which lasts for 18 months and

[1] Note that the ideal timing of milestones may have changed over the past 20 years and therefore may differ slightly for our sample from those depicted in Figures 2.1 and 2.2. In particular, the timing of the Warrant Officer Intermediate Level Course in our data (see Appendix A) does not match the timing shown in Figure 2.2, where that course lines up with the box labeled "PME" at 5 YAS.

FIGURE 2.1
Commissioned Officer Aviator Career Path

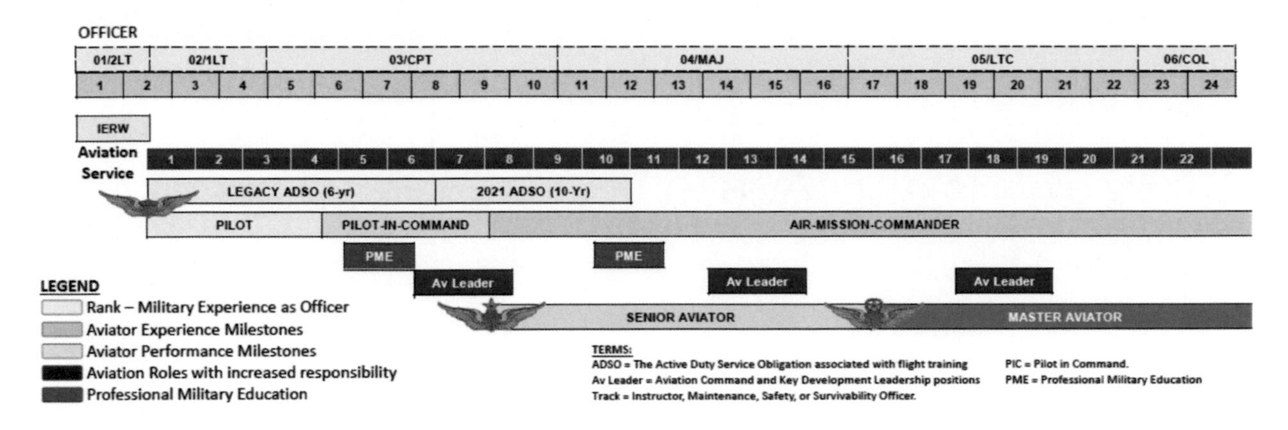

SOURCE: Created for the authors by the U.S. Army Aviation Center of Excellence, April 13, 2022.

FIGURE 2.2
Warrant Officer Aviator Career Path

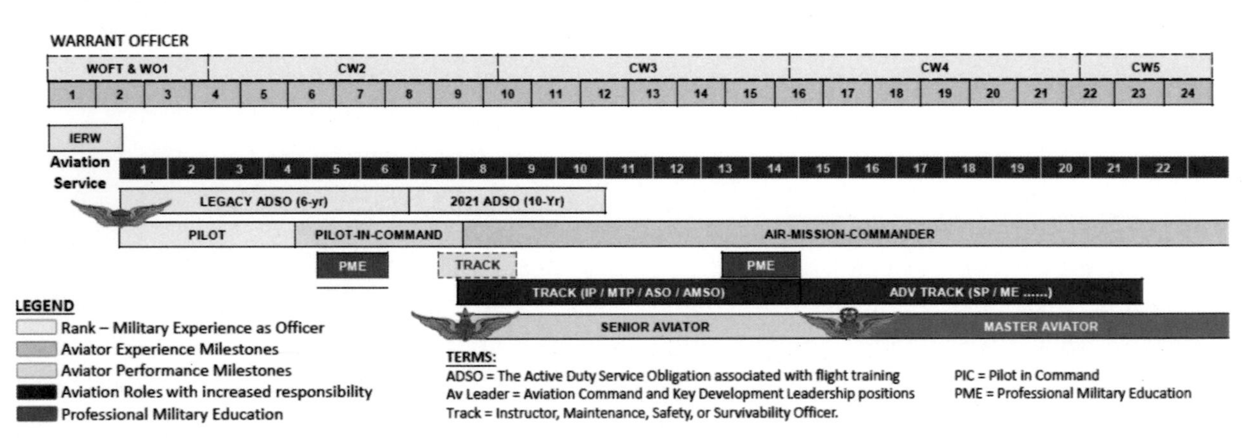

SOURCE: Created for the authors by the U.S. Army Aviation Center of Excellence, April 13, 2022.

grants the pilot aeronautical rating. Following graduation from flight school, both commissioned and warrant officers work toward earning their pilot-in-command rating. Ideally, pilot-in-command is achieved at 4.5 YAS. Both commissioned and warrant officers then attend PME at 5 YAS. For commissioned officers, the 5 YAS PME course is the Army Aviation Branch's Captain's Career Course, whereas the 5 YAS training for warrant officers is the Army Aviation Branch's Warrant Officer Intermediate Course. Shortly thereafter, ideally around 6 YAS, commissioned officers transition to command at the company level as aviation leaders, whereas warrant officers *track*, meaning that they specialize in one of four types of technical expertise (instructor pilot, maintenance test pilot [MTP], aviation safety officer, or aviation mission survivability officer). Both commissioned and warrant officers ideally achieve senior aviator around 8 YAS and master aviator around 16 YAS.

Achievement of milestones is likely based on multiple factors, revealed through discussions with our sponsor's office. One major factor is aviators' performance: higher performers will tend to achieve milestones faster, whether that is because they have a higher level of ability (therefore finding it easier to achieve milestones), because they put in more effort, or both. However, another potentially important factor is constraints and costs. For instance, achievement of pilot-in-command requires attaining a certain number of flight hours. Availability of opportunities to earn flight hours may depend on the unit an aviator is assigned to, the availability of aircraft, or other factors outside the officer's control. Other milestones may depend on the availability of seats in training courses. These factors should be kept in mind when modeling officers' attainment of career milestones.

Timing of Milestones in Army Data

Incorporating aviation milestones in the DRM required determining the probability that officers at a particular YAS will achieve each milestone. As a prerequisite step, we needed to identify milestone achievement for aviators in Army personnel data. We did this using data from the TAPDB, a repository of administrative personnel records from September of each year from 2002–2021 on commissioned and warrant officers in the military occupational specialties (MOSs) associated with the Army Aviation Branch.[2] In each year, we calculated the number of commissioned and warrant officers who had previously reached each milestone. We also calculated, within each cohort of entrants to the Army Aviation Branch, what percentage of entrants reached each milestone at each YAS. YAS is calculated using the difference between the file's calendar year and the year of the aviation service entry date provided in TAPDB; therefore, YAS is rounded to the nearest integer.

We calculated milestone achievement in TAPDB using the following procedures:

- Pilot status (IERW graduation): Our dataset includes aeronautical ratings by aircraft for multiple aircraft types.[3] Individuals are coded as having reached pilot status when they achieve qualified pilot status in at least one type of aircraft. If an individual is never observed as having achieved pilot status but is observed as having achieved a future milestone other than the Captain's Career Course, they are coded as having achieved pilot by the next milestone they are observed to have achieved.
- Pilot-in-command: Individuals are coded as having reached pilot-in-command when they achieve an aeronautical rating of pilot-in-command in at least one type of aircraft. If an individual is never observed as having achieved pilot-in-command but is observed as having achieved a future milestone other than the Captain's Career Course, they are coded as having achieved pilot by the next milestone they are observed to have achieved.
- Captain's Career Course (commissioned officers only): Commissioned officers are coded as having achieved the Captain's Career Course, which we use as a proxy for having taken command at the company level, when they are first observed as a graduate from the Captain's Career Course PME level.
- Warrant Officer Intermediate Course (warrant officers only): Warrant officers are coded as having achieved the Warrant Officer Intermediate Course when they are first observed as a graduate from the Warrant Officer Intermediate PME level or higher (i.e., senior service course or advanced course).
- Tracking (warrant officers only): Warrant officers are coded as having tracked when they are first observed as having been assigned a duty special qualifications identifier (SQI).[4]

---

[2] For commissioned officers, the relevant MOS codes (taken from the career field area of concentration variables in the TAPDB dataset) are as follows: aviation–general (15A), aviation combined arms operations (15B), aviation all-source intelligence (15C), aviation maintenance officer (15D), and aeromedical evacuation officer (67J). For warrant officers, the relevant MOS codes include the following MOS series: 150, 151, 152, 153, 154, and 155.

[3] We use the variables ACLEV01-ACLEV12 in TAPDB, which provide aeronautical ratings on each of 12 types of aircraft.

[4] Note that our data only include duty SQIs, not all SQIs that a warrant officer has achieved. We may therefore slightly undercount tracking, because tracked warrant officers may not be assigned to positions that require an SQI.

- Senior aviator: TAPDB includes information on over 40 awards for each individual. Individuals are coded as senior aviators the first time they are observed to have been awarded a combat and special skill senior army aviator badge or, if they are never observed to have been awarded a senior aviator badge, a combat and special skill master army aviator badge.
- Master aviator: Individuals are coded as master aviators the first time they are observed to have been awarded a combat and skill master army aviator badge.

Each of these statuses are cumulative; i.e., once achieved, they apply to all subsequent years of data for the individual.

We then examined the timing of each milestone among commissioned officers and warrant officers, for those who achieved each milestone. Appendix A provides more detail on the distribution of YAS at which each of these milestones is achieved. We used the timing to determine the probability that aviators would achieve the milestone, conditional on remaining in service long enough to achieve the milestone (with a grace period for some milestones to allow for variation in timing) and having achieved all previous milestones. These probabilities are used as inputs to the DRM.

Table 2.1 summarizes the ideal timing of each milestone as provided by our sponsor's office (and described in the previous section), the timing observed from our analysis of TAPDB, the timing used in the DRM, and the probability of milestone achievement, conditional on remaining in the military until the YAS of milestone achievement plus one to two grace years,[5] used in the DRM. For the most part, we chose timing in the DRM to correspond as closely as possible to the timing provided by our sponsor's office (in Table 2.1 this is marked as "Army input"). For instance, for pilot status, we chose 2 YAS because it is the closest YAS that we can observe in the DRM to 1.5 YAS, the timing of completion of IERW per Army input. We made one exception, which is the timing of the Captain's Career Course for commissioned officers. This exception reflects our use of the Captain's Career Course as a proxy for taking command at the company level, based on discussions with our sponsor's office on the appropriate milestones to include and how they are recorded in the data. We found that the Captain's Career Course is a prerequisite to taking command at the company level, and Army input suggested that taking command tends to occur at 6 YAS rather than 5 YAS. We also excluded the Warrant Officer Intermediate Course based on discrepancies between the timing suggested by the Army and the timing in the TAPDB data.[6]

---

[5] The grace years are included because, as shown in Appendix A, there is typically a window of three to four years during which a milestone is achieved (or at least where milestone achievement is recorded in TAPDB), and we wanted to capture everyone who achieved the milestones. The inclusion of grace years typically increases probabilities by 10–15 percentage points.

[6] Discussions with our sponsor suggested that the discrepancies in timing between the Army's input and the TAPDB were because of a recent policy change that redesigned the Warrant Officer Intermediate Course. Furthermore, we noticed substantial variation in the likelihood of completion of the course over time: Warrant officers were more likely to complete the course during periods where the Army was deliberately downsizing and less likely to complete the course during periods when the Army was expanding. Discussions with our sponsor revealed that the variation occurred because completion of the course was a determinant of being allowed to continue in the Army during periods of downsizing. In contrast, completion of the course was discouraged during periods where the Army did not have sufficiently high retention. This made determining a probability of completion for inclusion in the DRM very difficult, leading to its exclusion.

TABLE 2.1
Timing and Probability of Milestone Achievement

	Commissioned Officers				Warrant Officers			
Milestone	Milestone Timing (Army Input)	Milestone Timing (TAPDB)	Milestone Timing (Used in DRM)	Probability of Achievement (Used in DRM)[a]	Milestone Timing (Army Input)	Milestone Timing (TAPDB)	Milestone Timing (Used in DRM)	Probability of Achievement (Used in DRM)[a]
Pilot status	1.5 YAS	2–3 YAS	2 YAS	0.97	1.5 YAS	2–3 YAS	2 YAS	0.98
Pilot-in-command	4.5 YAS	3–4 YAS	4 YAS	0.65	4 YAS	3–4 YAS	4 YAS	0.91
First PME[b]	5 YAS	5–6 YAS	6 YAS	0.98	5 YAS	8–9 YAS	NA	NA
Track (WO only)	NA	NA	NA	NA	6 YAS	5–6 YAS	6 YAS	0.68
Senior aviator	8 YAS	8–9 YAS	8 YAS	0.56	8 YAS	10–12 YAS	10 YAS	0.65
Master aviator	16 YAS	16–18 YAS	16 YAS	0.24	16 YAS	16–18 YAS	16 YAS	0.54

NOTE: NA = not applicable; WO = warrant officer.

[a] Probability of achievement is the probability that an individual achieved the milestone at or before the milestone timing YAS plus two years, conditional on having stayed in the Army until the year of the milestone plus two years and having achieved all previous milestones (except pilots, whose probability was estimated at 2 YAS). The additional two years were added to account for the substantial variation in milestone timing documented in Appendix A. The 2012 to 2019 entering cohorts were used to calculate probabilities for pilot status, the 2012 to 2017 for pilot-in-command, the 2002 to 2015 cohorts for commissioned officers' first PME, the 2002 to 2013 cohorts for warrant officer tracking, the 2002 to 2012 cohorts for commissioned officers reaching senior aviator, the 2002 to 2010 cohorts for warrant officers reaching senior aviator, and the 2002 to 2014 cohorts for senior aviator. The changes in starting cohorts were made based on policy changes that affected the probability of making pilot or pilot-in-command earlier in an aviator's career; starting cohort does not strongly affect probabilities for the first PME or later milestones.

[b] First PME for commissioned officers is the Captain's Career Course. First PME for warrant officers is the Warrant Officer Intermediate Course.

Aviation Special and Incentive Pays

AvIP

We incorporated two S&I pays into the DRM: AvIP and AvB.[7] AvIP is a monthly special pay "intended to extend the aviation careers of Aviation Officers" (Army Regulation [AR] 600-105, 2020, p. 9). Officers who are entitled to basic pay, are in training for or maintain an aeronautical rating, achieve a minimum number of flight hours per month, are in aviation service for a specified period, and engage in regular performance of operational flying duty or proficiency flying duty are eligible to receive AvIP. After 12 YAS, aviators must meet additional requirements to maintain eligibility for AvIP, and commissioned officers cannot receive AvIP after reaching 25 YAS or being promoted higher than the rank of colonel. Warrant officers can continue to receive AvIP at the *over 10* (more than 10 YAS) level until retirement, and prior to January 2020, they could continue to receive AvIP at the *over 14* (more than 14 YAS) level until retirement (AR 600-105, 2020; Under Secretary of Defense for Personnel and Readiness, 2018).

A motivation for our project was Army concerns about the decline in the real value of AvIP and the fact that AvIP is guaranteed even to aviators who do not achieve key career milestones in a timely manner. AvIP was set at the values reported in Table 2.2 during FY 1998, which did not change until January 1, 2020, when the levels reported in Table 2.3 became effective (and, as of the writing of this report, AvIP has not changed since January 1, 2020). As a result, between FY 1998 and December 2019, the real value of AvIP declined substantially, as seen in Figure 2.3. The decline in the real value of AvIP could reduce retention and the ability

[7] AvIP was previously referred to as aviation career incentive pay (ACIP).

TABLE 2.2
Nominal Monthly AvIP, FY 1998–December 2019

YAS	Nominal Monthly AvIP ($)
2 or less	125
Over 2	156
Over 3	188
Over 4	206
Over 6	650
Over 14	840
Over 22	585
Over 23	495
Over 24	385
Over 25	250

SOURCE: Reproduced from DoDI 7000.14-R, 2001, Table 22-6.

TABLE 2.3
Nominal Monthly AvIP, January 2020–Present

YAS	Nominal Monthly AvIP ($)
2 or less	125
Over 2	200
Over 6	700
Over 10	1,000
Over 22	700
Over 24	400

SOURCE: Features data from DoDI 7000.14-R, 2022, Table 22-5.

of the Army to sustain its inventory of aviators, while guaranteed AvIP could mean that the Army continues providing the same retention incentive to aviators, regardless of milestone achievement or lack thereof.

AvB

AvB is a retention incentive for aviators with critical skills or in critical MOSs. In the Army, AvB is only available to warrant officers. More specifically, AvB is only available to warrant officers who have completed their ADSO or are within one year of completing their ADSO, who have 14 or fewer or 19 or more years of active federal service (AFS), who have less than six years of time-in-grade or are promotable, and who meet other requirements that vary year to year (U.S. Army Human Resources Command, 2022). AvB has been offered from time to time since FY 2000, the beginning of our study period, and according to discussions with our sponsor's office, the availability of AvB for an individual warrant officer is determined by overall retention within the Army Aviation Branch and by availability of funds. Appendix B describes the history of AvBs over the course of our study period. In FY 2020, tracked and fixed-wing warrant officer aviators in certain MOSs with less than 13 or between 19 and 22 years of AFS were eligible for $90,000 in AvB over a three-year contract (or, $30,000 per year for three years) (All Army Activities [ALARACT] 075/2019, 2019).

FIGURE 2.3
Real Value of AvIP, by YAS, FY 1998–FY 2019

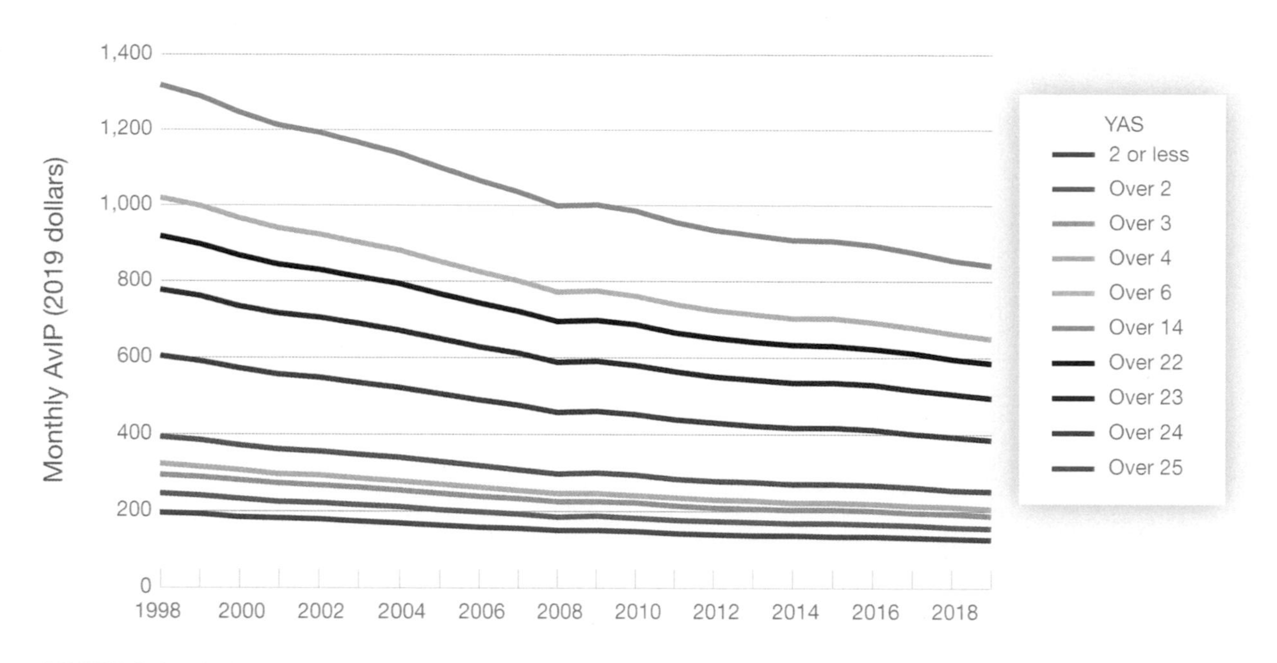

SOURCE: Authors' calculations using data from DoDI 7000.14-R, 2001; DoDI 7000.14-R, 2022; FRED, undated-a.

CHAPTER 3

Civilian Opportunities for Army Aviators

In the DRM, we modeled the decision of individual Army aviators to stay in the Army. In our model, that decision depended on the value of staying in the military versus leaving where the value of leaving was determined using civilian opportunities. Thus, in the model, the retention of Army officer aviators is, in part, determined by the civilian opportunities available to them. Aviators who leave the Army AC can choose to be a civilian pilot or pursue an alternative career. To help define the external demand for Army aviators, we gathered available information on the civilian market for Army aviators, including information about future civilian demand for fixed- and rotary-wing aviators, programs offering to retrain rotary-wing pilots as fixed-wing pilots, airline signing bonuses, and the structure of airline pay. This chapter summarizes our findings. Readers less interested in the details of the civilian market for pilots may wish to skip this chapter.

Much less is known about the civilian demand for rotary-wing or helicopter pilots versus fixed-wing pilots across the military services. Fixed-wing pilots in the military can transition to flying as civilian pilots for a major or regional airline. We found relatively little information about the demand for civilian helicopter pilots, as we discuss in this chapter. One interesting question is whether the demand for helicopter pilots is correlated with major airline hiring (MAH); i.e., is MAH a good instrument for measuring civilian demand for helicopter pilots or is there a better one? This chapter will explore and provide evidence, when available, of rotary-wing pilot demand and pay and compare them with fixed-wing pilot demand and pay.

One important consideration when an Army aviator is deciding whether to separate from the military is what they might earn in the civilian labor market. Thus, we conclude the chapter with a comparison of profiles of earnings by age of veteran pilots, veteran non-pilots, and the veteran population considered as a whole derived from the American Community Survey (ACS). We used the ACS profile of earnings by age in our retention models because these data specify earnings by age, whereas the data from other sources only give point estimates across the entire working-age population (that is, a single number for all ages). Pilot earnings show strong growth with age (possibly because of seniority rules governing compensation by major airlines), and a single point estimate would not reflect that growth.

Civilian Demand and Pay for Fixed-Wing Aviators

The civilian pilot shortage has been widely written about (see Greenberg, 2022; Schaper, 2022; and Baldanza, 2022 as examples), especially as it pertains to airlines and fluctuations in the civilian pilot job market during the coronavirus disease 2019 (COVID-19) pandemic. Airlines hired more each year between 2014 and 2020 when the COVID-19 pandemic hit (Future and Active Pilot Advisors [FAPA], undated). The increase in hiring was the result of a combination of factors, including (1) prior changes to the mandatory retirement age combined with the demographics of the civilian pilot workforce leading to a large number of pilots aging out of the workforce, and (2) growth in gross domestic product leading to systemwide growth in cargo and passenger miles, which in turn led to growth in the demand for pilots (Mattock et al., 2016). During that period of increased hiring, major airlines were seeking to hire as many military-trained fixed-wing pilots as possible. The introduction of more-stringent flight hour requirements by the Federal Aviation Administration (FAA) led to an increased demand for military-trained pilots, because they could qualify for a restricted airline transport pilot certificate with fewer total flight hours (750) than civilian-trained pilots (who would require 1,000 or 1,250, depending on their level of education) (Code of Federal Regulations, Title 14, Section 61.160). Research has shown a statistically significant correlation between MAH and U.S. Air Force pilot losses (McGee, 2015), more so for fixed-wing pilots who comprise the majority of Air Force pilots. As MAH has increased, the civilian demand for Air Force pilots has increased, which in turn generated more Air Force pilot losses. The same concern was expressed by the U.S. Navy in the Office of the Under Secretary of Defense for Personnel and Readiness' report (2019, p. 7), that the correlation between increased MAH and post-command resignations was affecting military (and Navy) retention.

Civilian Demand for Fixed-Wing Aviators

Figure 3.1 shows the trend in MAH since 2014. MAH has increased since at least 2014, except for the COVID-19 outbreak in 2020. And, as an outlier, the hiring in 2022 exceeded any previous year by the end of May 2022. This likely occurred because airlines needed time to get their training programs back up and running at full capacity. Furthermore, pilots that were furloughed and needed to return likely needed to become current and qualified in the aircraft again to meet safety standards, and many pilots that were near retirement age were offered lucrative contracts to retire early during the pandemic, leaving a larger hole to fill (Joseph, 2020).

In 2021, an update to results published in McGee (2015) and Terry et al. (2019) forecasted an unprecedented level of civilian pilot hiring. The forecasted level is exceptionally high in 2022 at 10,000 and is expected to remain above 5,000 for the next ten years starting in 2023, as shown in Figure 3.2. The data shown in Figure 3.2 came from a model run from November 2021 and was updated based on FAPA data (FAPA, undated). The forecasted hiring for the next ten years is expected to be at levels higher than hiring seen in every year prior except 2021. Therefore, based on McGee (2015) and Terry et al. (2019), we expect unprecedented demand for military-trained pilots and a highly likely increase in military pilot losses as a

FIGURE 3.1
Historical Major Airline Hiring Over Time

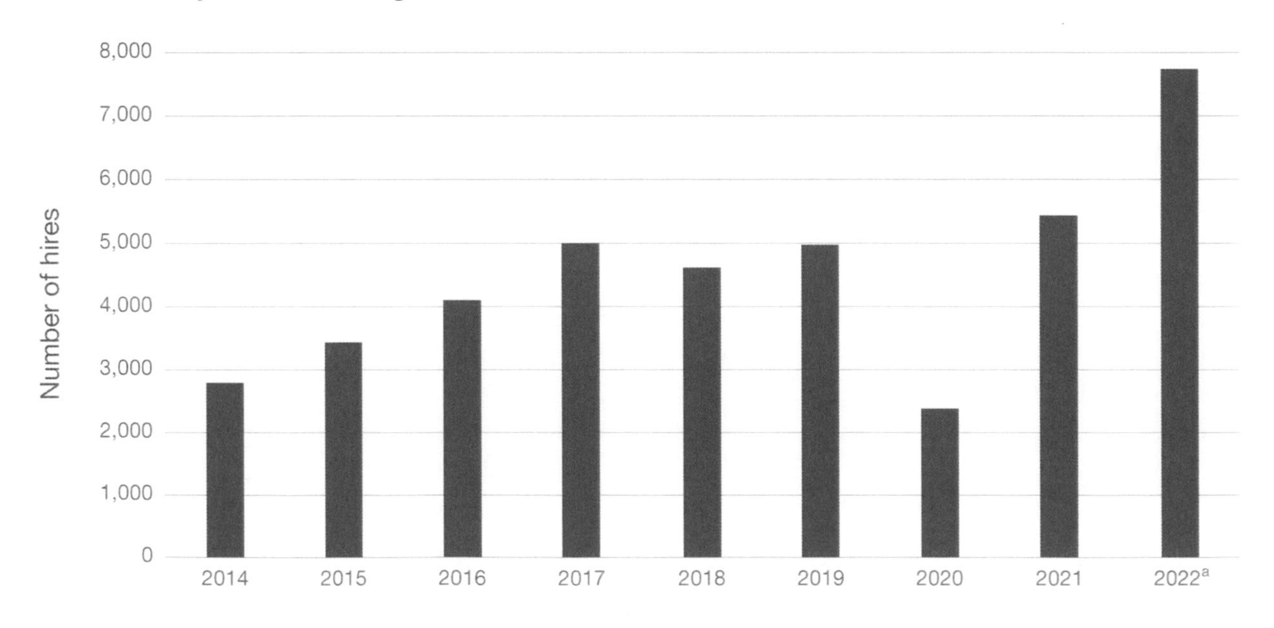

SOURCE: Features data from Future and Active Pilot Advisors, undated (with 12 major airlines).
[a] The 2022 hiring numbers are for January–July 2022 only.

FIGURE 3.2
Historical and Forecasted Major Airline Hiring

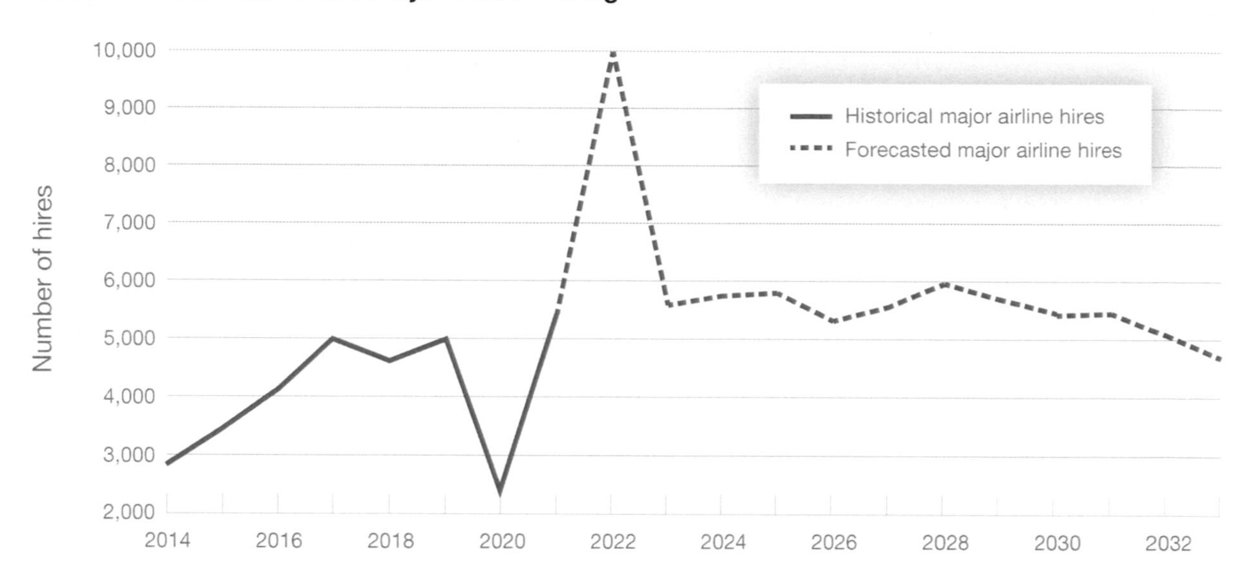

SOURCES: Features historical data from FAPA, undated; and forecasted estimates based on a model documented in McGee (2015) and Terry et al. (2019), and authors' analysis of publicly available data from FAPA, undated, the FAA Aerospace Forecast Fiscal Years 2021–2041 (FAA, 2021), and airline financial submissions (Southwest, 2021; Delta Air Lines, 2021; United, 2021; American Airlines, 2021; FedEx, 2021).

result. This demand is likely to extend to rotary-wing pilots as well given the recent emergence of rotary to fixed-wing transition programs, which we will discuss in more detail later.

Civilian Pay for Fixed-Wing Aviators

Because previous RAND research presents information on major airline pay as of 2014 (Mattock et al., 2016), we focused here on how monetary compensation has changed since then. Table 3.1 compares average hourly pay in 2014 from Mattock et al. (2016), with hourly pay in 2022 for *Year 12 captains* (civilian pilots with 12 years of employment with a given airline) for wide-body aircraft.

The table shows that for five of the 12 major airlines included in the earlier tabulations,[1] pay increased for captains with 12 years of employment with a given airline. The real median increase in pay was 6.6 percent. This is representative of how the pay for pilots has increased across the board as the demand for pilots in the civilian sector has increased. Table 3.2 presents some salary information for several airlines at different years and seniority levels. The relative increase in pay enhances the incentive for a service member to leave the military if they are interested in becoming a commercial pilot.[2]

Effective annual first-year pay at regional airlines has increased, up from an average of $28,523 (2022 dollars) in 2014 (McGee, 2015) to more than $50,000 in 2022, as shown in Table 3.3. The creation of signing bonuses and tuition reimbursement is evidence of greater demand. This greater demand may be due, in part, to regional airlines being a feeder for the major airlines' pilot pipeline (Joseph, 2022b).

[1] The 12 major airlines are defined by FAPA, undated.

[2] As we discuss in Chapter 4, the DRM does not compare civilian pay with military pay at each point in the career to judge competitiveness; rather it compares the value of staying with the value of leaving where the value of staying includes current and (discounted) future compensation, including expected retired pay, a stochastic factor, and a factor that represents non-monetary factors, such as taste.

TABLE 3.1
Comparison of 2014 and 2022 Hourly Pay for Year 12 Captains of Wide-Body Aircraft for a Select Number of Airlines

Airline	2014 Hourly Pay for Captain in Year 12 of Wide-Body Aircraft (2014 dollars)	2014 Hourly Pay for Captain in Year 12 of Wide-Body Aircraft (2022 dollars)	2022 Hourly Pay for Captain in Year 12 of Wide-Body Aircraft (2022 dollars)	Real Increase
American	233	291	342	17.3%
Delta	255	319	340	6.6%
FedEx	254	318	326	2.6%
United	255	319	352	10.3%
United Parcel Service (UPS)	255	319	340	6.6%

SOURCES: Features 2014 data from Mattock et al., 2016; 2022 data from Airline Pilot Central, undated.

TABLE 3.2
Estimated Pay for Different Positions and Years

Airline	First Officer (Year 1)	Captain (Year 6)	Captain (Year 12)
United	$93,820	$282,400	$295,660
Delta	$100,840	$285,460	$298,720
American	$98,800	$289,540	$305,860
FedEx	$86,560	$270,160	$284,440
UPS	$55,960	$317,080	$331,360
Southwest	$92,680	$268,120	$286,480

SOURCE: Features data from ATP, undated.

NOTE: Estimates are based on flying a 737 for 85 hours per month and an additional $7,000 annual per diem.

As a result of the increase in demand for civilian fixed-wing pilots at regional and major airlines, as well as the resulting increase in pay and additional benefits, employment as a commercial airline pilot may become increasingly attractive for service members considering leaving active duty.

Civilian Demand and Pay for Rotary-Wing Aviators

Unfortunately, publicly available wage data do not distinguish between rotary-wing and fixed-wing pilots. However, the distinction may not be as important as it might seem at first glance. In particular, numerous rotary to fixed-wing transition programs have emerged, which suggests that MAH is a reasonable indicator of the civilian demand for helicopter pilots. Such programs are likely to continue and expand in response to the ongoing surge in MAH in 2022.

Civilian Demand for Rotary-Wing Pilots

When researching civilian demand for helicopter pilots, a few sources can be found (Crookston, 2021; AVweb, 2019; Aviation Technician Education Council, 2019; Gordon, 2018). Forecasts of future helicopter pilot demand are not as complete as those provided in and before 2018. Boeing's forecast of helicopter pilot

TABLE 3.3

Signing and Retention Programs and First-Year Pay at Selected Regional Airlines

Airline	Signing and Retention Program	Effective Annual First-Year Pay
Air Wisconsin	$5,000 training completion bonus	$52,191
CommutAir	$5,000 airline transport pilot tuition reimbursement as certified flight instructor and $6,000 retention bonus; direct-entry Captains receive up to an $18,000 signing bonus	$50,330
Endeavor	$10,000 training completion bonus	$72,065
Envoy Air	$15,000 sign-on bonus	$74,174
GoJet	No signing or retention program	$47,129
Horizon Air	$12,500 flight training stipend paid $5,000 after instrument rating and $7,500 after commercial multiengine land	$53,545
Mesa	$5,000 airline transport pilot tuition reimbursement as certified flight instructor and $7,500 retention bonus	$53,772
Piedmont	$7,500 cadet bonus and $15,000 sign-on bonus	$72,130
PSA	$15,000 sign-on bonus	$74,378
Republic	$5,000 airline transport pilot tuition reimbursement as certified flight instructor and $6,000 retention bonus drawn from $17,000 sign-on bonus	$70,737
SkyWest	$17,500 airline transport pilot tuition reimbursement as certified flight instructor; $7,500 bonus for pilots with current turbojet type rating is available upon completion of initial operating experience	$56,123

SOURCE: Features data from ATP, undated.

demand over the next two decades fell from 59,000 to 44,000 between 2018 and 2019, but subsequent forecasts did not separate out the demands for helicopter pilots from those of fixed-wing aircraft pilots. Moreover, the emergence of rotary to fixed-wing transition programs render the distinction between rotary-wing and fixed-wing pilots less important (AVweb, 2019; Aviation Technician Education Council, 2019; Boeing, 2020; Boeing, 2021).

Honing into available evidence on helicopter pilot demand, a source from University of North Dakota (UND), in collaboration with Helicopter Association International and Helicopter Foundation International (Gordon, 2018), predicted a shortage of 7,649 helicopter pilots in the United States between 2018 and 2036. Assuming the methodologies were consistent between the UND forecast and the Boeing forecasts, the U.S. helicopter pilot demand is 13–17 percent of the worldwide forecasted demand.

The research conducted in the 2017–2018 time frame suggested that helicopter pilots were in demand and that demand would continue for the next 20 years.

Retraining Rotary-Wing Pilots to Be Fixed-Wing Pilots

Regional air carriers are another potential source of demand for fixed-wing pilots. Regionals hire newer or inexperienced pilots, who then gain experience and flight hours flying the regional routes for lesser pay, and once the pilots have reached the requisite certifications, experience, and flight hours, the pilots often try to get hired by a major airline at a higher rate of pay. Thus, regional air carriers have a higher rate of turnover as pilots move on to the majors as soon as they are able (Joseph, 2022a).

This demand has resulted in regionals looking for sources of pilots elsewhere. Prior to the COVID-19 pandemic, regional air carriers were starting to create and run rotary to fixed-wing pilot transition programs. UND reached out to three regional airlines to determine interest:

> "Lo and behold, the interest in these transition programs is very high. In 2017 alone, 500 [rotary] pilots transitioned through their programs with a 95 percent completion rate. That was at just three of these airlines. Many more are offering these programs." (Gordon, 2018)

In 2022, the world seemed to begin transitioning from a COVID-19 pandemic response to an endemic way of life. Many aspects of life—including domestic and foreign travel—resumed, resulting in MAH forecasting an even greater need for pilots. This shift should have ramifications on helicopter pilot demand as well. It would be logical to assume that rotary to fixed-wing transition programs would resume and flourish. However, trying to find definitive sources beyond anecdotes proved difficult. There are three programs we came across in our research that are involved in rotary to fixed-wing transition programs.

The Infinity Flight Group is located at the Trenton-Mercer Airport in Trenton, New Jersey. It provides a rotary transition program (RTP) in partnership with Mercer County Community College (Infinity Flight Group, undated). The RTP is designed to train commercially rated helicopter pilots how to fly commercial, fixed-wing aircraft in the fastest time possible and at little to no cost. They work with rotary-wing pilots to help determine the best funding sources to minimize the cost, including using Veteran's Affairs benefits. The Infinity Flight Group, in partnership with regional airline Piedmont Airlines, offers a guaranteed progression from Piedmont Airlines to American Airlines. Candidates can reach out through the website to receive more information about the program.

The Veteran to Aviation Charity (VAC) (formerly the Rotary to Airline Group) is a 501(c)3 nonprofit founded by four former Army aviators with the initial intention of addressing the national pilot shortage by training both professional and experienced rotary-wing aviators (Rotary to Airline Group, undated). The group's purpose has since evolved to placing veterans in the airline industry, regardless of experience. VAC uses RTP to help helicopter pilots successfully transition to roles as commercial fixed-wing pilots. VAC has developed several scholarships to aid transitioning veterans. In a call with founder Erik Sabiston, he said that VAC is starting to look toward chief warrant officers and enlisted ranks as the next source of personnel to feed this pipeline and continue to feed the demand for commercial airline pilots.[3]

One RTP known to have existed prior to the pandemic is Envoy Air. Envoy's RTP aims to train military rotary-wing pilots to fly commercial fixed-wing aircraft and earn a restricted airline transport pilot certificate (Envoy, undated). Envoy has partnered with ATP Flight School to provide the training and the training plan and costs. The ATP Flight School helps interested pilots gain experience while providing opportunities for tuition reimbursement, financial assistance, and benefits through ATP's 38 airline hiring partnerships. Envoy's website states that it is not accepting RTP candidates currently; therefore, we do not know if the program continues to operate.

Civilian Pay for Rotary-Wing Pilots

We attempted but struggled to find data on earnings for helicopter pilots as a function of age or experience similar to that available for pilots of fixed-wing aircraft (e.g., Table 4.1. "Hourly Pay Rates for Selected Airlines, 2014" in Mattock et al., 2016).[4] There is some information from the U.S. Bureau of Labor Statistics in its Occupational Employment and Wages from May 2021 under the category of 53-2012 Commercial Pilots. The description of commercial pilots is as follows: "Pilot and navigate the flight of fixed-wing aircraft on nonscheduled air carrier routes, or helicopters. Requires Commercial Pilot certificate. Includes charter pilots with similar certification, and air ambulance and air tour pilots. Excludes regional, national, and interna-

---

[3] Erik Sabiston, founder, Rotary to Airline Group, interview with the authors, December 29, 2021.

[4] Current pay rates for airlines can be found on Airline Pilot Central, undated.

TABLE 3.4
Percentile Wage Estimates for Commercial Pilots

Percentile	10%	25%	50% (Median)	75%	90%
Annual Wage	$50,080	$75,370	$99,640	$134,110	$205,940

SOURCE: Features data from U.S. Bureau of Labor Statistics, 2022.

tional airline pilots" (U.S. Bureau of Labor Statistics, 2022). The mean wage estimate for commercial pilots is $115,080. The percentile wage estimates for commercial pilots are provided in Table 3.4.

The uncertainty of how much of the data in Table 3.4 is from helicopter pilots versus fixed-wing pilots makes it difficult to come to any conclusions regarding the distribution of wages for helicopter pilots. Our research uncovered some other sources, but they lack the rigor and specificity needed for this project. An Everyday Aviation website post on helicopter pilot salary reports an overall average salary for a helicopter pilot with 12 years of experience of $93,883, averaged across all the industries, and a starting salary for new helicopter pilots of $30,000–$40,000 per year (Everyday Aviation, undated). Once 1,000 hours of flight time is reached, helicopter pilots could earn $50,000–$90,000 per year depending on what industry they are in and how much experience they have. Some examples include emergency medical services helicopter pilots who average $65,000 annually and tour pilots who average $50,000–$60,000 annually. Helicopter pilots with over ten years of experience could earn up to $120,000 per year depending on the industry they serve and their position.

Salary.com shows the average helicopter pilot salary in the United States is $101,192 as of August 29, 2022, but the range typically falls between $85,820 and $130,105 (Salary.com, 2022b). Salary ranges can vary widely depending on many factors, including education, certifications, and experience levels. The average chief helicopter pilot salary in the United States is $165,290 as of August 29, 2022, but the range typically falls between $146,028 and $206,605 (Salary.com, 2022a). Another similar type of source, Talent.com, shows the average helicopter pilot salary in the United States is $69,430 per year. Entry-level positions start at $40,000 per year while most experienced workers make up to $95,324 per year based on 460 salaries (Talent.com, undated), showing lower average pay levels than Salary.com.

While these data give some indication to the range of salaries that could be expected, they do not include sufficient information to calculate how salaries grow with experience, thus making it infeasible to calculate the lifetime career value of a helicopter pilot. The potential for helicopter pilots to transition to higher-paying fixed-wing aircraft positions further complicates the picture. Given the high level of demand for fixed-wing pilots and the increased salary and benefits, it is likely that the MAH demand has increased the demand for all military-trained pilots, including Army aviators. Army aviators have the choice to remain rotary-wing pilots with the same opportunities that have existed for many years (law enforcement, ambulatory, tourism, etc.) and the additional opportunity to become fixed-wing pilots. With the existence of rotary to fixed-wing transition programs, this is now a more realistic alternative, one that did not exist more than five years ago. Therefore, we hypothesized that civilian commercial demand for fixed-wing pilots is a reasonable proxy for the demand for helicopter pilots.

Civilian Earnings Data for Pilots and Non-Pilots, by Education Level

One important consideration when an Army aviator is deciding whether to separate is what they might earn in the civilian labor market. We included this in our models by constructing a profile of earnings by age that represents the opportunity wage. This section shows data derived from the ACS on the earnings of the individual populations of pilots and non-pilots, as well as the two populations combined. We distinguished between the earnings of those individuals with some college (one to three years) and those with four years of college, because there is a substantial premium associated with being a college graduate. We used statistics on the earnings of veterans with some college to model the opportunity wage of warrant officers, while we used the statistics on the earnings of veterans with four or more years of college to model the opportunity wage for commissioned officers. These are admittedly imperfect approximations of the true opportunity wages. For example, some warrant officers have attended four years of college either on separation or sometime thereafter. It is also not uncommon for commissioned officers to pursue graduate degrees.

Our interest in understanding the earnings of non-pilots as well as pilots is because pilots who leave military service do not necessarily choose a civilian occupation that involves flying (Lankford, 2017). Following our earlier work (Mattock et al., 2016), we initially investigated DRMs of Air Force pilots where the opportunity wage facing an individual is a weighted sum of a fixed percentile of the non-pilot and pilot wage trajectories associated with their level of education, where the weighting is a function of the level of MAH. When preliminary analyses showed that the level of MAH was not a significant factor in the DRMs, we also investigated models using a fixed percentile of the earnings of the combined population of pilots and non-pilots.

Earnings of the Combined Population, by Education Level

In this section, we show the earnings for the combined population of veteran pilots and non-pilots by education level, which were used as inputs for the final version of the DRM. Then, as supporting detail, we show the earnings of veteran pilots and non-pilots for individuals with some college and for veterans with four years of college, which were used as inputs to the alternative version of the DRM shown in Appendix E. Finally, we make some concluding remarks.

The 50th percentile of earnings of veterans with one to three years of college peaks at approximately \$63,000, as can be seen in Figure 3.3. By comparison, the 50th percentile of earnings of veterans with four years of college peaks at approximately \$88,000, showing a *college premium*—the additional monetary benefits pilots with more college receive—of \$25,000, as shown in Figure 3.4. In both cases, earnings peak between ages 50 and 55.

Earnings of Veteran Pilots and Non-Pilots with One to Three Years of College

Figure 3.5 shows the profile of earnings by age for non-pilot veterans with one to three years of college, with earnings peaking between ages 50 and 55. As mentioned, these curves are relevant for measuring the civilian opportunities for warrant officer aviators. The curves were calculated using a Tobit model as described in Appendix C. Figure 3.6 shows the earnings profiles for pilot veterans. Notice that pilot earnings are substantially higher than non-pilot earnings at each percentile. For example, the peak earnings at the 50th percentile for non-pilots is just under \$65,000, while the peak earnings at the 50th percentile for pilots is nearly \$140,000, more than double the earnings of non-pilots. Another difference is that pilot earnings peak between ages 55 and 60, while non-pilot earnings peak between ages 50 and 55.

Earnings of Veteran Pilots and Non-Pilots with Four Years of College

The earnings of veterans with four years of college, the group relevant for commissioned officers in our analysis, show a similar pattern to veterans with some college, in that pilots show substantially greater earnings than non-pilots. In addition, the earnings of veterans in both categories with four years of college are greater than those of veterans with one to three years of college. Figure 3.7 shows the earnings of non-pilots with four years of college. The 50th percentile of earnings peaks between ages 50 and 55 at just under \$90,000, approximately \$25,000 more than for comparable non-pilot veterans with one to three years of college. The

FIGURE 3.3
Earnings for Veterans with One to Three Years of College

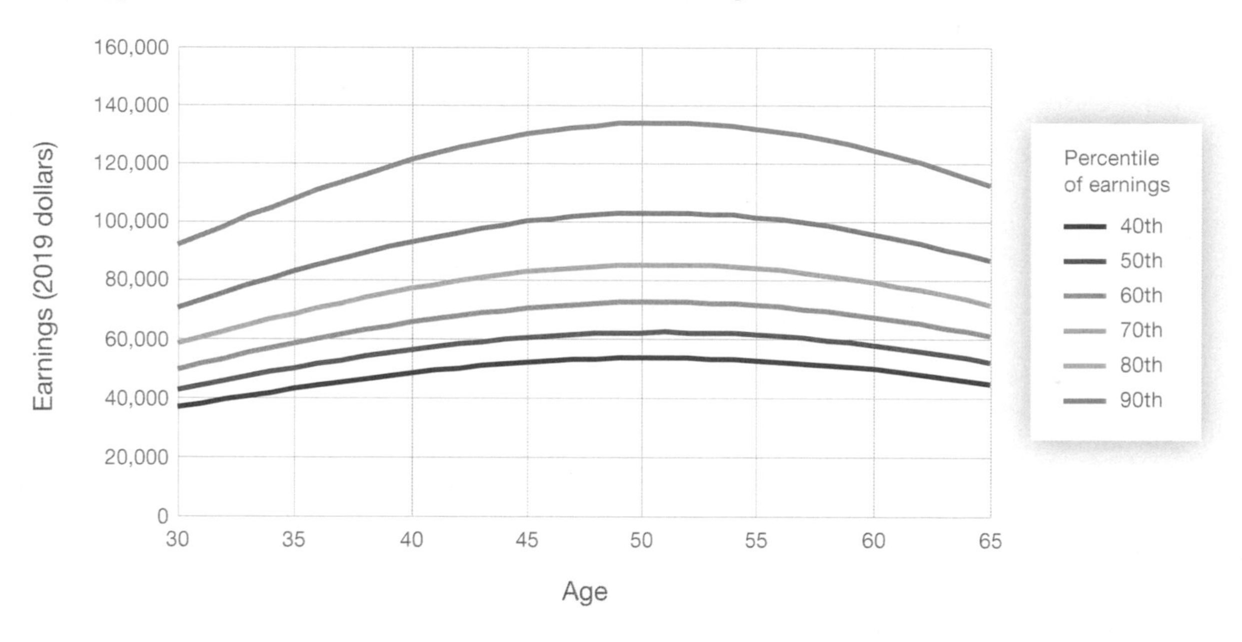

SOURCE: Authors' calculations of data from Ruggles et al., 2021.

FIGURE 3.4
Earnings for Veterans with Four Years of College

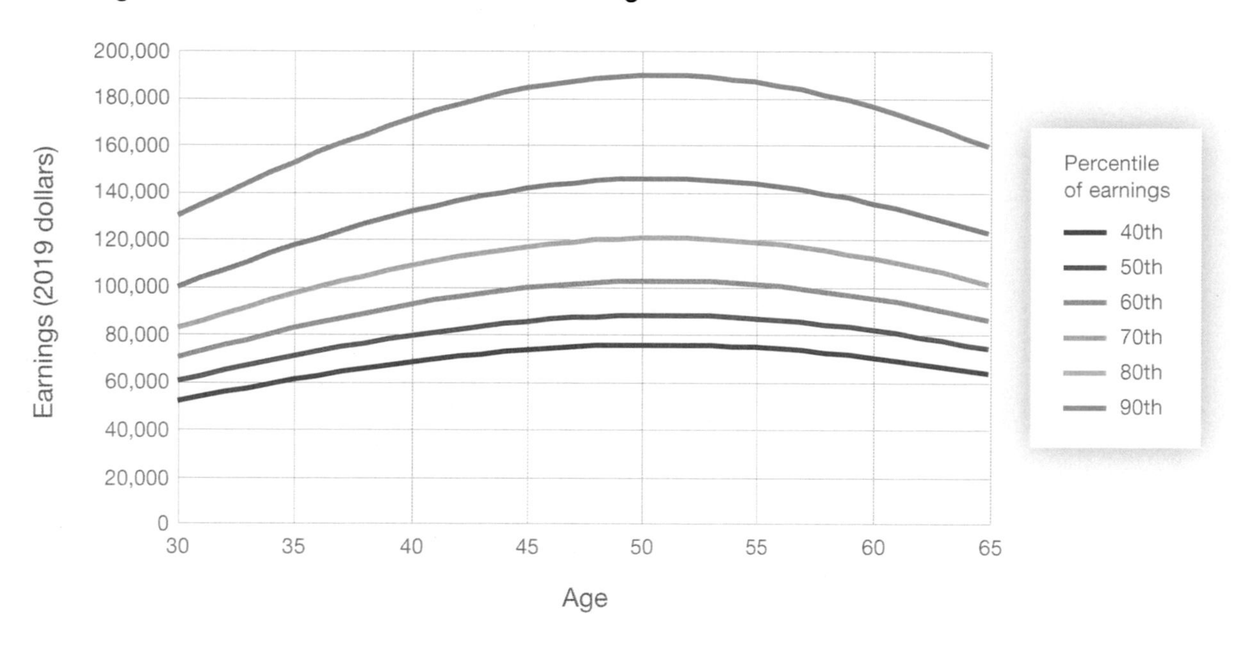

SOURCE: Authors' calculations of data from Ruggles et al., 2021.

FIGURE 3.5
Earnings for Non-Pilots with One to Three Years of College

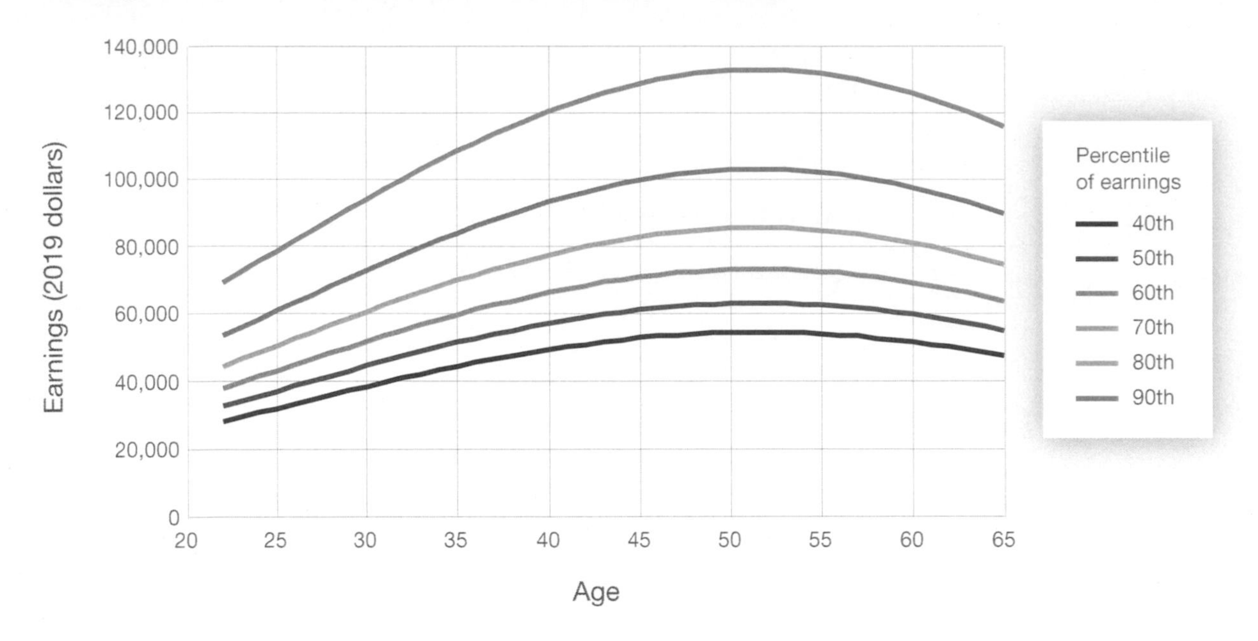

SOURCE: Authors' calculations of ACS data (U.S. Census Bureau, undated).

FIGURE 3.6
Earnings for Pilots with One to Three Years of College

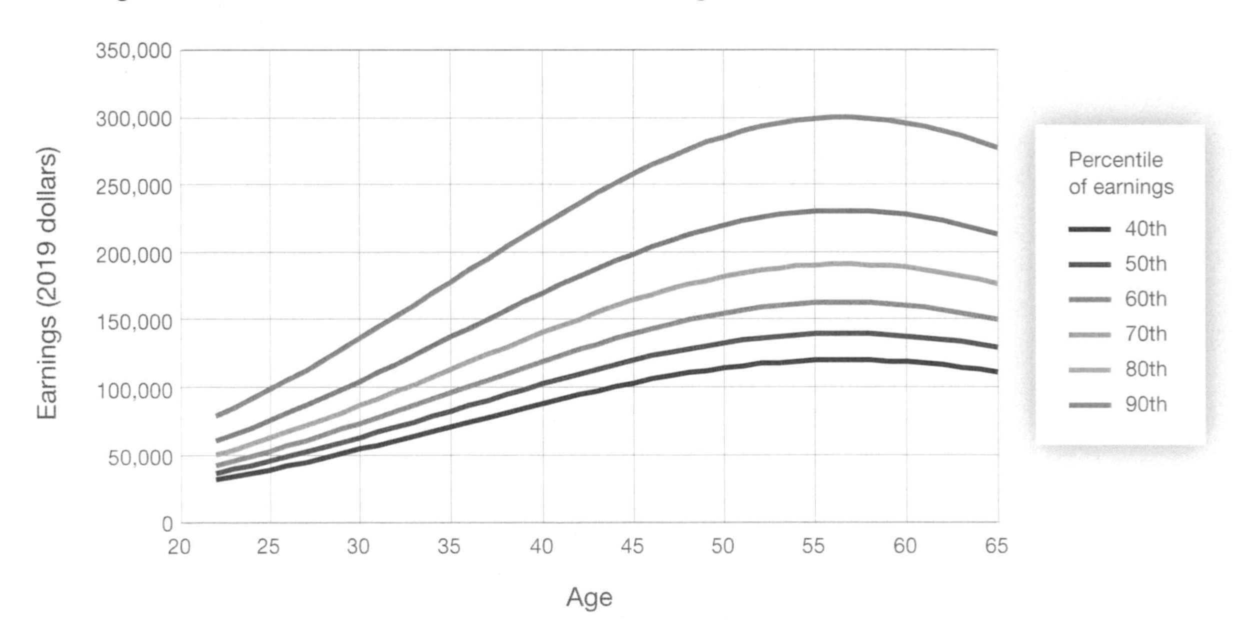

SOURCE: Authors' calculations of ACS data (U.S. Census Bureau, undated).

earnings of pilot veterans are substantially higher, as can be seen in Figure 3.8, with the 50th percentile of earnings peaking between ages 55 and 60 at slightly less than $175,000—$85,000 more than the earnings of non-pilots and approximately $35,000 more than the earnings of pilot veterans with one to three years of college. Similar to veterans with one to three years of college, pilot earnings peak five years later than those of non-pilots, with peak earnings for pilots occurring between ages 55 and 60 while peak earnings for non-pilots occur between ages 50 and 55. The later peak earnings for pilots are likely driven by the strict seniority-based compensation system seen in many airlines.

Summary

The demand for civilian pilots has risen recently and is anticipated to remain strong in future years. Pilot earnings tend to be higher than those of non-pilots, and so the measures of opportunity cost for civilians as a whole at any given percentile could possibly understate the opportunity cost for military aviators. Although existing evidence of salaries of rotary-wing pilots suggests that their earnings are lower than those of fixed-wing pilots, the emergence of rotary to fixed-wing transition programs means that the opportunity cost of rotary-wing military aviators could potentially be better approximated by the earnings of fixed-wing civilian pilots.

Despite this, as we show in Appendix E, in our retention model (where the opportunity wage facing an individual is a weighted sum of a fixed percentile of the non-pilot and pilot wage trajectories associated with their level of education, and the weighting is a function of the level of MAH), we found that MAH did not play a large role in the retention of Army aviators, especially warrant officers. So, we instead estimated models using a fixed percentile of the earnings of the combined population of pilots and non-pilots, the 50th percentile for commissioned officers and the 70th percentile for warrant officers, as shown in Figure 3.9. Note that the two lines are close in value and show much the same curvature. Their closeness may account for the

FIGURE 3.7
Earnings for Non-Pilots with Four Years of College

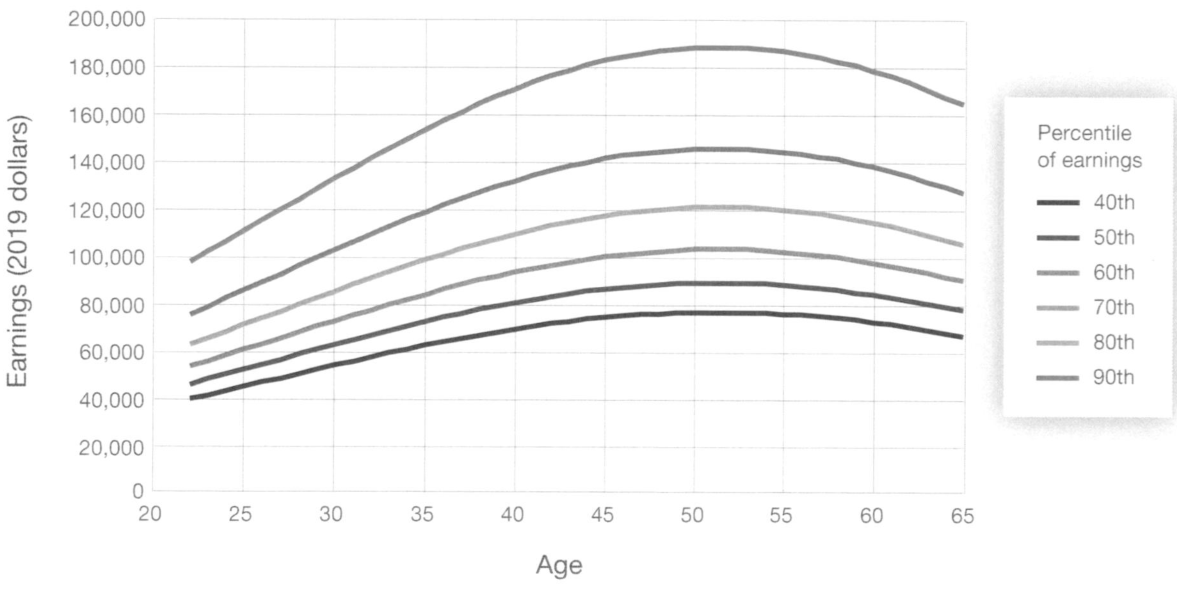

SOURCE: Authors' calculations of ACS data (U.S. Census Bureau, undated).

difference in the estimated taste parameters for commissioned officers and warrant officers that we discuss in the following chapter, which is similar in size to the college premium we observed earlier in this chapter.

FIGURE 3.8
Earnings for Pilots with Four Years of College

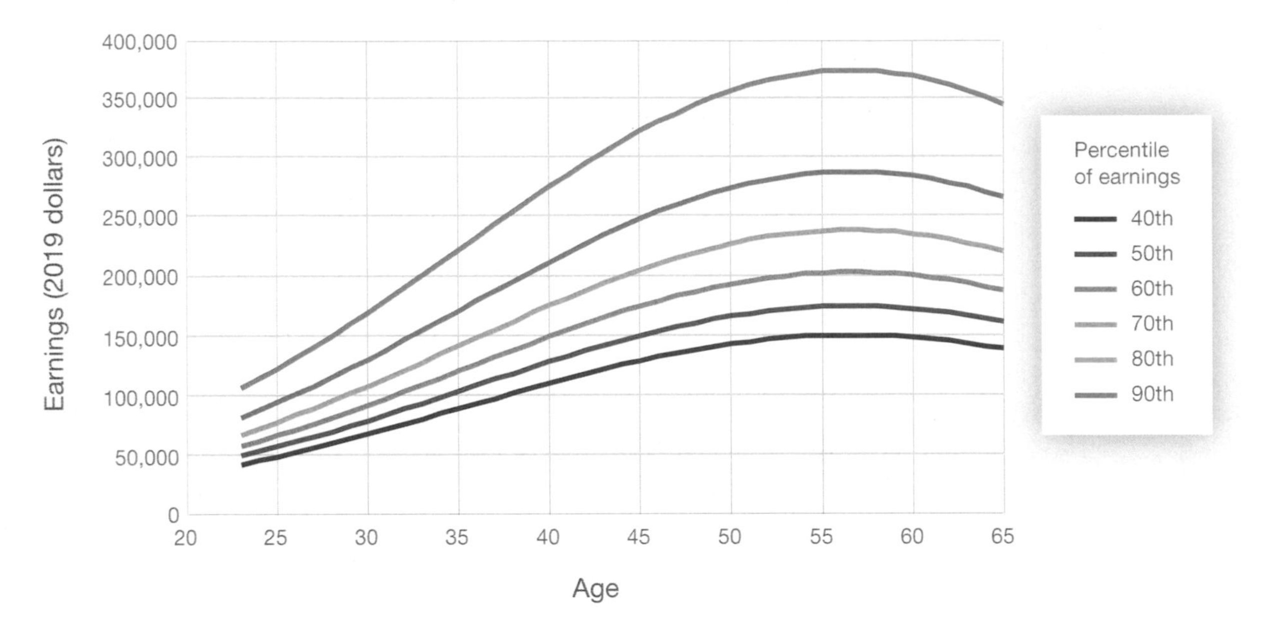

SOURCE: Authors' calculations of ACS data (U.S. Census Bureau, undated).

FIGURE 3.9
Earnings for Veterans Used in Dynamic Retention Models

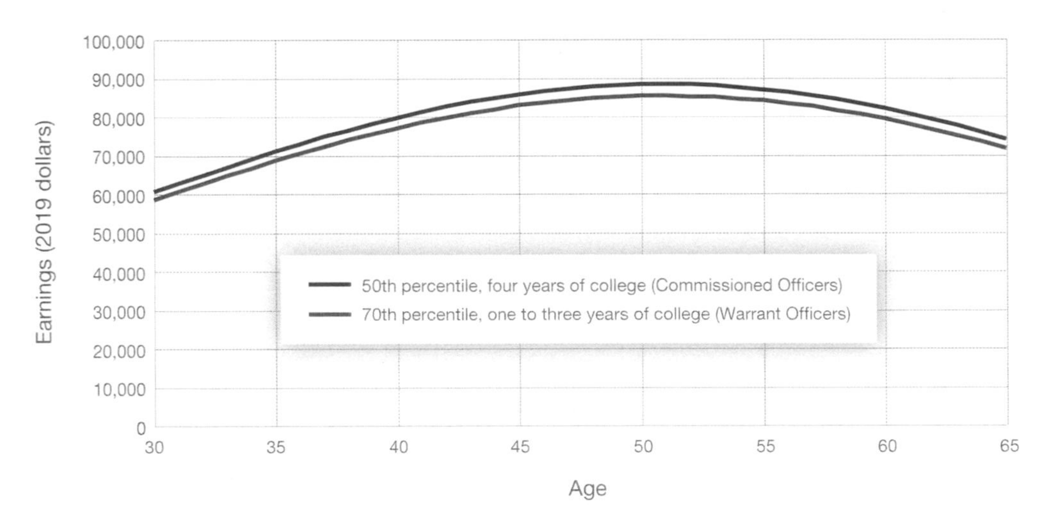

SOURCE: Authors' calculations of data from Ruggles et al., 2021.

CHAPTER 4

DRM Approach and Model Estimates

This chapter provides a technical description of the specification of the DRM used in our analysis of compensation in the Army Aviation Branch. The DRM is a dynamic, stochastic model of each service member's decision to stay or leave the Army. For each period, we calculated each aviator's value of staying in or leaving the Army, conditional on model parameters. In each period, each aviator chooses to stay if the value of staying is higher than the value of leaving, and vice versa. We used the observed behavior of aviators to calculate the model parameters. Our simulations then used the model parameters and the structure of the model to predict behavior under compensation changes.

This chapter first describes in detail the structure of the DRM and estimation methodology. It then presents the model estimates, discusses model fit, and shows how the assumptions used are relevant to the simulations. Detailed discussion of how we used the model estimates to simulate alternative policies is contained in the next chapter. Readers not interested in such technical details are invited to proceed to the discussion of model fit at the end of this chapter or to skip the chapter entirely. This chapter draws heavily on Mattock et al. (2016) and Asch et al. (2020).

DRM Overview

The DRM is a stochastic dynamic programming model of the decision to stay or leave military service. The model begins when individuals enter active military service. In each period, an individual can choose to continue in or leave active service. Those who leave are assumed to work in the civilian labor market. We assumed that those who depart active service never return, and we ignored the possibility of participation in a reserve component, because we do not have TAPDB data on reserve component participation for our entire sample period (typically, DRM models also include the option to be a reservist, participating in either the U.S. Army Reserve (USAR) or the Army National Guard (ARNG)). The decision at each point in time of whether to continue in active service or to leave active duty is a function of whether they are currently active, their total years of AFS, their taste for active service, and unknown factors that are treated as random shocks from a known statistical distribution.

Formally, the value of staying active at time *t* is equal to

$$V_t^S = V_t^A + \varepsilon_t^A,$$

where V_t^A is the non-stochastic value of the active alternative, and ε_t^A is the random shock at time *t*.[1] Similarly, the value of leaving at time *t* is

$$V_t^L = V_t^C + \varepsilon_t^C,$$

[1] To simplify the notation, we omit a subscript *i* to denote the individual aviator.

where V_t^C is the non-stochastic value of being a civilian, and ε_t^C is the random shock at time *t*. The shock terms represent factors that are unobservable to researchers but inevitably arise and affect the decision of whether to stay or leave active duty. Such factors include but are not necessarily limited to the individual's assessment of the quality of their assignment; mission danger; civilian opportunities not captured by our wage and hiring measures; opportunities for promotion; the desirability and availability of duty locations; a change in marital status, dependency status, or health status; or the prospect of deployment or deployment itself. The individual is assumed to know the distributions that generate the shocks and to know the shock realizations in the current period, but not in future periods. The distribution of the shocks is assumed to be constant over time, and the shocks are assumed to be serially uncorrelated. At each point in time, the individual chooses the path with the highest expected value, which can differ between individuals with the same non-stochastic values V_t^A and V_t^C because of the random shocks. Next period, each individual reoptimizes. Even if the non-stochastic values were to remain constant over time, some individuals who remained in active service last period will choose to depart this period because of new draws of the random shocks.

We assumed that the shocks have extreme value distributions. The extreme value distribution, denoted as *EV*, is characterized by a location parameter *a* and a scale parameter *b*. The mean equals $a + b\phi$, where ϕ is Euler's gamma (~0.577), and the variance equals $\pi^2 b^2/6$. As we derived in past studies (see, for example, Asch et al., 2008), this implies that

$$\varepsilon_t^A, \varepsilon_t^C \sim EV[-\phi\kappa, \kappa],$$

where κ is the scale parameter of the distribution of ε_t^A and ε_t^C.

The value of the alternatives, V_t^A and V_t^C, depend on the current-period pay associated with serving in the AC or working as a civilian, W_t^A or W_t^C. W_t^A contains regular military compensation (RMC) and any period-specific S&I pays, including AvIP. RMC is the sum of a service member's basic pay, their allowances for housing and subsistence, and the value of the tax advantage associated with being able to receive these allowances tax-free as provided in Table B-4 of the Green Books (OUSD P&R, 2000–2019). In addition, the value of the alternatives also includes each individual's taste for active duty, γ^A. The taste term, γ^A, represents the net advantage of being in the AC relative to being a civilian. Previous estimates of the average taste for active duty among individuals without prior active service were negative, suggesting that military pay must compensate for the hardships and risks associated with service, although estimates for commissioned officers with prior enlisted service have found positive mean values of γ^A (Asch et al., 2018). All else equal, a higher taste for active duty increases active retention. These tastes are assumed to be constant for each individual over time but may vary across individuals. We do not observe γ^A; rather, we assume that γ^A has a normal distribution among AC entrants and estimate the mean (μ^A) and standard deviation (σ^A) of the distribution.

The non-stochastic value of staying active can therefore be written as

$$V_t^A = \gamma^A + W_t^A + \beta E\left[\max\left[V_{t+1}^S, V_{t+1}^L\right]\right],$$

where β is the personal discount factor.

The expected value of the best choice in the next period, $E\left[\max\left[V_{t+1}^S, V_{t+1}^L\right]\right]$, is a key feature of dynamic programming models that distinguishes them from other dynamic models. The presence of this term means that individuals are able to reoptimize; that is, they are able to select a different alternative in the future when new information becomes available. In the current period *t*, with future realizations of the shocks unknown, the best the individual can do is to estimate the expected value of the best choice in the next period—i.e., the expected value of the maximum. This is true in the next period, and the one after it, and so on. Thus, individuals in the model are forward-looking and rationally incorporate the possibility of future career changes and future uncertainty into current decisions.

The non-stochastic value of the civilian choice is the present discounted value of all future civilian compensation, written as

$$V_t^C = \sum_{s=t}^{T} \beta^{s-1} W_t^C + R_t,$$

where W_t^C is expected civilian pay (described in more detail in Chapter 3, Appendix C, and later in this chapter). R_t in the civilian equation is the present value of any active military retirement benefit for which the individual is eligible, and T is the final year of the model. We modeled R_t using the legacy retirement system because most career decisions for individuals in our data were made under the legacy retirement system; the BRS only became effective in January 2018 and covered all new entrants as of that date and those in service as of December 31, 2017 who were eligible to opt into the new system and who chose to do so.[2] As noted above, we assumed that once an individual has left the AC they cannot reenter and therefore remain a civilian for the remainder of the sample period.

Similar to previous studies using the DRM (e.g., Mattock et al., 2016), the model also incorporates a *switching cost*. This term reflects the cost of leaving active duty before the initial ADSO is completed, because leaving before one's obligation is complete is more difficult than doing so at the completion of one's ADSO. This cost is not actually paid by the individual but is implicit in making certain transitions.

The individual recognizes that today's choice affects military and civilian compensation in future periods. Although the individual does not know when future military promotions will occur, they do know the promotion policy and can form an expectation of military pay in future periods. Similarly, the individual can form expectations of future civilian pay.

Extensions of the DRM for Warrant Officers

The description so far forms the basis for our specification of the DRM for commissioned officers. However, the DRM must be extended for warrant officers to include the possibility of three-year contracts and enlisted service prior to aviation service. As pay varies by aviation milestone achievement for warrant officers, but not commissioned officers, the DRM estimation must also be extended to include the achievement of aviation milestones for warrant officers, and the simulation capability must be extended for commissioned officers to allow us to simulate the effects of basing special pay on aviation milestones for them as well. The next sections discuss these modifications.

Including AvB

Over the period covered by our data, Army warrant officer pilots were eligible for multi-year (typically three-year) contracts under which they would be paid an AvB retention bonus that typically increased with the length of the service commitment that the individual elected. As mentioned earlier, the availability of contracts and the rules governing eligibility for these multi-year contracts varied over time. Consequently, following Mattock and Arkes (2007) and Mattock et al. (2016), we extended the DRM to include the AvB by adding equations that express the value of the AvB program for different ADSOs. The DRM described above involves two equations: the first is the value of staying active, while the second is the value of leaving. The extension to include AvB changes only the value of staying active, without changing the value of leaving.

The equation $V_t^{S/1}$ gives the value of staying active for one additional year, at time *t*. Thus, we can write the value of staying active for one more year as

[2] All aviators included in the estimation of our model entered the Army under the legacy retirement system and thus would only have made decisions under the BRS if they chose to opt in. Our decision to model retirement benefits using the legacy system only is based on evidence that few service members chose to opt into the BRS (Philpott, 2019).

$$V_t^{S/1} = V_t^{A/1} + \varepsilon_t^{A/1} = \gamma^A + W_t^{A/1} + \beta E\left[\max\left[V_{t+1}^S, V_{t+1}^L\right]\right] + \varepsilon_t^{A/1},$$

where $W_t^{A/1}$ includes RMC and AvIP. Similarly, we can write the value of staying active and taking the AvB with a three-year obligation as

$$V_t^{S/3} = V_t^{A/3} + \varepsilon_t^{A/3} = \sum_{n=0}^{2}\beta^n\left[\gamma^A + W_{t+n}^{A/3}\right] + \beta^3 E\left[\max\left[V_{t+3}^S, V_{t+3}^L\right]\right] + \varepsilon_t^{A/3},$$

where $W_t^{A/3}$ includes RMC, AvIP, and AvB for a three-year contract. If a pilot is not eligible for a three-year contract, $V_t^{S/3} = -\infty$. We denote $V_t^S = \max\left[V_t^{S/1}, V_t^{S/3}\right]$.

An eligible pilot compares the value of leaving, V_t^L, with the maximum of the value of staying active for one year, $V_t^{S/1}$, and three years, $V_t^{S/3}$. If the three-year option is offered and the pilot is eligible, the probability that an initially offered pilot stays active is

$$\Pr\left(V_t^S > V_t^L\right) = \Pr\left(\max\left[V_t^{S/1}, V_t^{S/3}\right] > V_t^L\right).$$

We modeled the contract length choice as a nested choice that incorporates the fact that the decision to stay for one more year versus three more years is nested within the larger choice of staying in general. That is, individuals decide whether to stay in the Army or leave, and then conditional on staying, they decide whether to stay for one year or (if they are eligible) stay for three years and receive AvB. We assumed that the random shocks of the contract length choice follow an extreme value distribution, so that $\varepsilon_t^{A/1}, \varepsilon_t^{A/3} \sim EV\left[-\phi\lambda, \lambda\right]$ where λ is a second shape parameter (one more than in the model for officers). We also redefined the scale parameter for the distribution of ε_t^A and ε_t^C to be τ. The scale parameter in the value function for staying in general with the contract length choice nested within this larger choice is then modified to be $\kappa = \sqrt{\lambda^2 + \tau^2}$. When estimating the model, we estimated κ and λ and treated τ as slack.

We assumed that members have perfect foresight over future available contract choices and used the eligibility requirements for AvB to constrain the choice set as needed.

Including Prior Enlisted Service

We followed Asch et al. (2018) in extending the DRM for warrant officers to include individuals with prior enlisted service who make up a large portion of our sample of warrant officers.[3] To do so, we allowed total YAS to be less than an individual's total years of active service as a warrant officer.[4]

As before, in considering each alternative, the individual accounts for their state and type, with years of enlisted service and YOS as a warrant officer taking the place of years of AFS. Years of enlisted service factor into the calculation of retirement benefits, pay, and (in some years) eligibility for AvB.[5]

Including Aviation Milestones

We incorporated aviation milestones into the model in a manner similar to the way that we have previously incorporated promotion to higher grades (see, for example, Asch et al., 2020). The model for warrant officers must account for milestones because warrant officer S&I pay (in particular, AvB) is currently contingent on

---

[3] We did not include commissioned officers with prior enlisted service because they are relatively rare.

[4] While the DRM could be adapted to model the decision to pursue a warrant officer career, we did not do so here because that would significantly expand the project beyond the time and resources available. We instead focused on the decision to stay in or leave active service following the transition to service as a warrant officer.

[5] Observing the patterns in our data, we believe that the TAPDB records for warrant officers with prior enlisted service are missing time spent in IERW, whereas the records for warrant officers without prior enlisted service seem to include that time. Therefore, for the purposes of the estimation, we added two years to the number of years of enlisted service for all warrant officers with any enlisted service.

their achievement. Milestones are currently not implemented in officer pay or promotion. For this reason, milestones are not incorporated into the estimation but are incorporated in the simulations for officers as described in Chapter 5.

As described in Asch et al. (2020), we assumed that the timing and probability of milestone achievement at each YAS is the same across all commissioned officers and is the same across all warrant officers. The timing and probabilities are shown in Chapter 2. Variation in the timing and probability of milestone achievement for an individual service member is captured by the shock term. Achievement of a particular milestone occurs at a given number of YAS, but the probability of achievement differs by milestone. Also, the probability of milestone achievement is assumed to be invariant to policy changes that could affect the constraints on officers' ability to achieve milestones. Failure to reach a milestone can decrease the value of continuing in the military and therefore decrease retention. For warrant officers, this is true at baseline because milestones affect AvB eligibility. For commissioned officers, we later found this is true in the simulations.

Expected Civilian Pay

A crucial input to the DRM is the *outside option*, or the civilian earnings that an individual would expect when considering exiting the AC. We constructed estimates of expected civilian earnings using the ACS as described in Chapter 3 and Appendix D (Ruggles et al., 2022). Our ACS sample is limited to full-time, full-year workers. We also limited our sample to individuals with at least some college to match the education requirements for Army warrant officers.

We used the ACS data to construct estimates of civilian earnings percentiles by age and education level, as described in Appendix C. We constructed W_t^C for commissioned officers using the 50th percentile of civilian earnings by age for college graduates, shown in Figure 3.8, and we constructed W_t^C for warrant officers using the 70th percentile of civilian earnings by age for individuals with some college, shown in Figure 3.7.[6] W_t^C is in 2019 dollars, scaled up using the Consumer Price Index for Urban Consumers (CPI-U; FRED, undated-a), and also incorporates the yearly unemployment rate, multiplying the earnings percentile by 1 minus the yearly unemployment rate (FRED, undated-b).

Estimating the DRM

For commissioned officers, we estimated four model parameters:

- The mean and standard deviation of the taste for active service relative to civilian opportunities (μ^A and σ^A). We assumed a normal distribution for active tastes.
- A scale parameter reflecting the dispersion of the shock affecting the active and civilian states (κ). We assumed an extreme value distribution for the shock.
- A switching cost that is incurred if the individual leaves active duty before completing ADSO that is a linear function of the number of years that the individual has remaining in ADSO.

We estimated the same four model parameters for warrant officers, plus one additional scale parameter to incorporate the choice of contract term length (λ). We assumed a discount factor of 0.94 for commis-

[6] The choice of different percentiles for commissioned and warrant officers reflects that these percentiles produced the best model fit. We note that throughout the project, we found that it was the *curvature* of the earnings percentile, rather than the level, that mattered most for fit. Choosing an earnings percentile with incorrect curvature pushed the estimate of σ^A close to zero in some cases. Once the correct curvature was found, any issues with the level could be corrected by μ^A.

sioned officers, following Mattock et al. (2016), which is consistent with previous estimates of the discount factor for commissioned officers (e.g., Mattock, Hosek, and Asch, 2012).[7] We also assumed a discount factor of 0.91 (halfway between the value used for commissioned officers and the value we previously estimated or assumed for enlisted, 0.88; see, e.g., Mattock, Hosek, and Asch, 2012; Asch et al., 2020).

The model was estimated using data from TAPDB on career decisions (that is, presence in or absence from the Army's inventory of current active-duty soldiers) made by individuals who entered aviation service between 2002 and 2010, tracking these decisions through 2021. Prior service for warrant officers was calculated as the total number of YOS recorded in the enlisted file, taken from the TAPDB records from those warrant officers' enlisted service from the year before they became warrant officers.[8] Our data include a total of 2,785 commissioned officers and 4,117 warrant officers.

The key information to be gleaned from the model is the probability of choosing a particular alternative, active or civilian, called the *transition probability*. Because the shocks are assumed to have extreme value distributions, we were able to derive closed-form logistic expressions for each transition probability. For example, the probability that a commissioned officer chooses to stay active at time *t*, given that they are already in the AC and are eligible to stay one year, is given by

$$\Pr(V_t^S > V_t^L) = \frac{\exp\left(\frac{V_t^A}{K}\right)}{\exp\left(\frac{V_t^A}{K}\right) + \exp\left(\frac{V_t^C}{K}\right)}.$$

If an individual is a civilian at time *t*, the transition probability that that individual chooses to remain a civilian at time $t + 1$ is set equal to 1. Because, by assumption, the transition probabilities in different periods are independent of one another when an individual is free to make a choice—that is, the model is a Markov decision process—the transition probabilities for each period can be multiplied to obtain the probability of any given career profile of active and civilian alternatives that we observe in the data. Multiplying the career profile probabilities together gives an expression for the sample likelihood that we used to estimate the model parameters for each occupation using maximum likelihood methods.

The transition probability for a warrant officer who is eligible for AvB and who has been eligible for AvB in previous years is slightly more complicated. Service members may have the option to make multiple contract choices over their careers. For example, they might initially choose a one-year contract, followed by a three-year contract, and then separate. Unfortunately, the information available to us in TAPDB does not include contract choice, so observing that a service member is in the AC in year *t* may mean that they chose to stay for one year during year *t*, chose to stay for three years during year *t*, or are in the second or third year of a three-year contract during year *t*. We therefore calculated the probability of observing a pilot staying a particular number of years and then leaving or being censored (i.e., the data end before the pilot leaves) in the follow-

[7] Assumptions on the value of the discount factor may affect our simulation results. Too high a discount factor would lead to individuals placing too high a value on future pay relative to current pay, which would lead to larger anticipatory effects of changes in compensation (e.g., if AvB increases for 19–22 years of AFS, then a high discount rate would lead to a large retention increase at 17 and 18 years AFS, because they want to stay in the Army long enough to receive AvB). Too low a discount factor would lead to individuals placing too low a value on future pay relative to current pay, which would lead to lower anticipatory effects of changes in compensation but potentially larger effects of changes in current compensation.

[8] Previous versions of the DRM have typically used the Work Experience file from the Defense Manpower Data Center (DMDC) (e.g., Mattock et al., 2016). However, the DMDC data do not include information on aviation milestones, which was crucial for implementing the model that includes aviation milestones. Early versions of the model included in this report were estimated using DMDC data, both with and without reserve participation, and produced similar model estimates and fit graphs.

ing manner. We first constructed a set of all sequences of contract lengths that the service member has been eligible for in their career up until year *t* and then calculated the cumulative probability of each sequence. Following Mattock and Arkes (2007) and Mattock et al. (2016), we exploited the fact that many paths have a near-zero probability, noting that if one term of a product of probabilities is 0, the entire expression is 0. This assumption saves us from having to explicitly calculate the other terms in the cumulative probability expression. We then summed up the probability of all possible sequences of contract decisions to calculate the cumulative probability of remaining active until time *t*. We repeated the process to calculate the cumulative probability of remaining active until time *t-1* and then calculated the transition probability as

$$\Pr\left(V_t^S > V_t^L\right) = \frac{\text{CumPrActive}_t}{\text{CumPrActive}_{t-1}}.$$

We estimated the model using maximum simulated likelihood methods where optimization is done using the Broyden-Fletcher-Goldfarb-Shanno algorithm, a standard hill-climbing method. Random draws for taste for active service were made using a Halton sequence with 101 integration points. Standard errors of the estimates were computed using numerical differentiation of the likelihood function and taking the square root of the absolute value of the diagonal of the inverse of the Hessian matrix. To judge goodness of fit, we used the parameter estimates to simulate retention rates by YOS of personnel and compared those rates with the actual data. We show goodness-of-fit diagrams in the next section, where we present the models' parameter estimates.

Once we had parameter estimates, we then used the logic of the model and the estimated parameters to simulate the AC cumulative probability of retention in each YOS in the steady state for a given policy environment. These policies include an increase in AvB, a change in the level of AvIP, or an increase in the civilian opportunity wage facing Army pilots. By *steady state*, we mean when the force consists solely of service members who have spent their entire careers under the policy environment being considered. The simulation output includes a graph of the AC retention profile by YOS. We show model fits in the next section by simulating the steady-state retention profile in the baseline—or current policy environment—and comparing the simulation with the retention profile observed in the data.

Parameter Estimates and Model Fit

For ease of estimation, all parameters are estimated in logs. Tables 4.1 and 4.2 present parameter estimates, standard errors, z-statistics, transformed values, and descriptions of each parameter for commissioned officers and warrant officers, respectively. All coefficient values have the expected sign and are in the range of estimates we have observed in previous work.

To test the fit of our model, we ran Monte Carlo simulations of the retention behavior of samples of commissioned officers and warrant officers whose distributions of calendar year of entry to aviation service (and for warrant officers, the joint distribution of calendar year of entry to aviation service and years of enlisted service) match the distribution of the observed sample. We then compared the results of the simulation with the observed retention profile of Army aviators.

The findings are summarized in Figure 4.1. The black solid lines are the Kaplan-Meier survival curves for the observed sample, representing the proportion of the observed sample who remained in the AC at each year since entering the AC (or for warrant officers with prior enlisted service, since entering the AC as a warrant officer). The black dashed lines show 95-percent confidence intervals for the observed sample. Note that we only observe individuals who completed IERW. The red lines are the Kaplan-Meier survival curves for the simulated sample. The key finding is that the model fits the data well for both commissioned officers

TABLE 4.1
Parameter Estimates for Commissioned Officers

Variable	Parameter Estimate	Standard Error	z-statistic	Transformed Value	Description
ln(Kappa)	6.1854	0.2655	23.2985	485.6088	Scale parameter
-ln(Mu)	4.5199	0.0448	100.9326	−91.8221	Mean taste for active service
ln(SD)	4.3052	0.6550	6.5725	74.0808	Standard deviation of taste for active service
−ln(Switch)	6.0500	0.3073	19.6871	-424.1041	Cost of leaving active service before end of initial ADSO
beta	0.9400	—	—	—	Assumed discount factor
Likelihood	−6,064.8500				
Observations	2,785.00				

TABLE 4.2
Parameter Estimates for Warrant Officers

Variable	Parameter Estimate	Standard Error	z-statistic	Transformed Value	Description
ln(Kappa)	5.0124	0.0393	127.4815	150.2700	Scale parameter for between nests
ln(Lambda)	3.5684	0.1522	23.4514	35.4611	Scale parameter within nest
-ln(Mu)	4.0911	0.0204	200.6395	−59.8047	Mean taste for active service
ln(SD)	2.2120	0.4774	4.6332	9.1338	Standard deviation of taste for active service
-ln(Switch)	4.6308	0.0411	112.5776	−102.6002	Cost of leaving active service before end of initial ADSO
beta	0.9100	—	—	—	Assumed discount factor
Likelihood	−9,468.5700				
Observations	4,117.00				

(left panel) and warrant officers (right panel) for most YOS, with the model underpredicting commissioned officer retention in years 5 and 6 and slightly overpredicting retention in years 7 through 12.

Summary

In this chapter, we discussed the basic structure of the DRM and modifications we made to allow the achievement of aviation milestones, for warrant officers to receive AvB, and for warrant officers to have prior enlisted service. We found that the model fits well, especially for warrant officers. Our model estimates suggest that warrant officers have a slightly stronger taste for active service than commissioned officers, likely related to

FIGURE 4.1

Predicted Active Retention Compared with Observed Retention for Commissioned Officers and Warrant Officers Entering Service Between 2002 and 2010

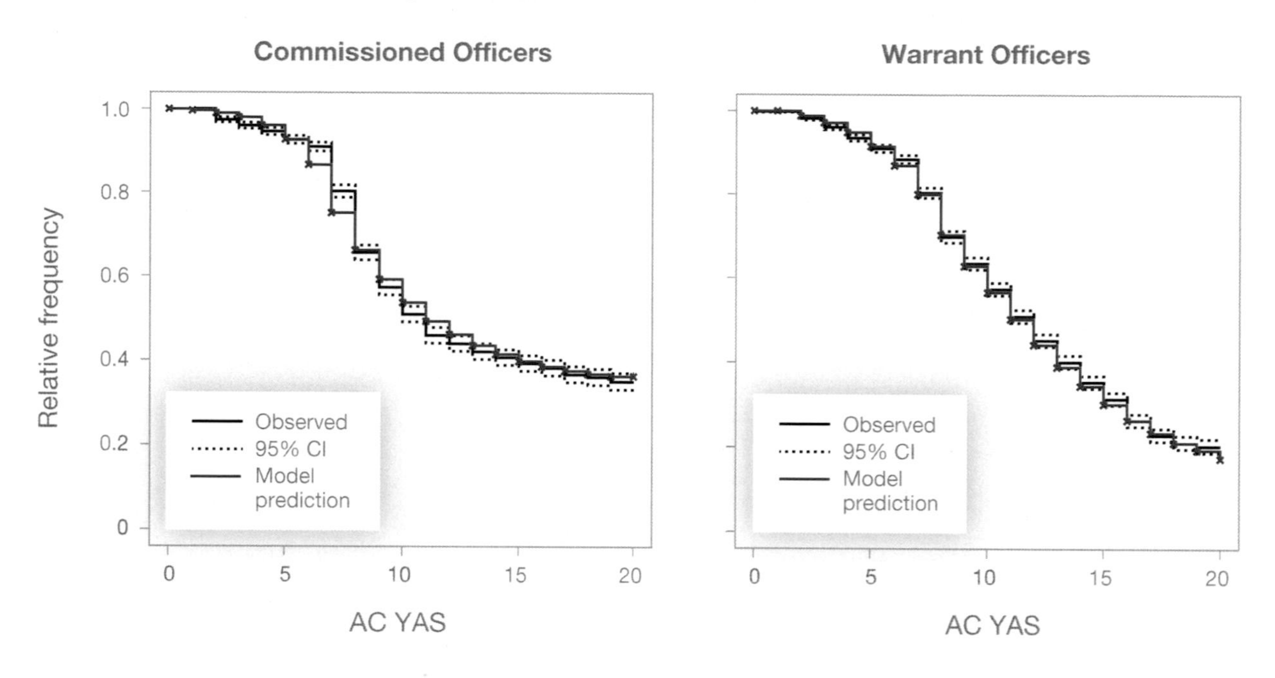

NOTE: CI = confidence interval.

the fact that many warrant officers have prior enlisted service (and therefore would be positively selected on taste for active service), but commissioned officers have a larger standard deviation of taste for active service and a higher cost of leaving the Army prior to the end of their initial ADSO.

The next chapter goes into our simulation results.

CHAPTER 5

Simulated Retention Effects of Alternative Special and Incentive Pays

This chapter presents results of simulations of the retention effects of alternative S&I pays for Army aviators. We present results with varying values of AvIP, including a change in the way AvIP is determined so that it depends solely on aviation milestones, as well as with varying values of AvB.

Interpreting Simulation Results

All simulations should be interpreted as representing a *steady state*; that is, they represent the retention profile of a single entering cohort over its entire career, if the cohort spent its entire career under a particular incentive pay scheme. A simulation in the steady state can give policymakers information necessary to predict the full trajectory of retention over the career of aviators who have recently entered the Army, which is especially helpful when setting AvIP in a way that will optimize the inventory of qualified aviators going forward.

Unless otherwise stated, all simulations assumed (1) a ten-year ADSO, (2) AvIP set to the values in effect as of January 1, 2020, (3) no AvB for warrant officers, and (4) a probability of reaching each milestone equal to the value observed in our data. The assumption of a ten-year ADSO is potentially problematic. Our models were estimated using individuals who entered Army Aviation when the initial ADSO was six years, and all simulations assumed that the mix of individuals who enter Army Aviation does not respond to simulated policy changes. This assumption is likely reasonable for most of the changes in AvIP and AvB that we simulated in this report but seems unlikely in the case where the requirement for staying on active duty rises by four years. Instead, we believe that the mix of individuals who entered Army Aviation after the change in initial ADSO length is likely to have had a stronger taste for the Army on average (that is, more positive tastes, on average), and therefore may have had different retention behavior. We therefore included simulation results under a six-year initial ADSO in Appendix D to explore whether our results are sensitive to the assumption regarding initial ADSO length.

We also incorporated a cost estimation capability into the simulations, because cost is an important factor in the decision to adopt a particular AvIP scheme. AvIP cost is calculated by adding up AvIP paid to every individual who remains in the Army at each YAS and should be interpreted as the cost of providing AvIP to a cohort over the course of a potential 30-year career. Total cost is calculated as the sum of the military or RMC pay bill of the aviator inventory, AvIP costs, and the retirement accrual charge the Army faces for the aviator force. Specifically, it is the sum of RMC paid to every individual who remains in the Army at each YAS, 38.9 percent of the basic pay bill at each YAS (to account for the cost of retirement under the legacy retirement system; see DoD Office of the Actuary, 2020), and AvIP cost. We reported only percentage changes in cost as the cost itself depends on the entering cohort size. Cost may change for two reasons. First, different AvIP values will change AvIP costs for a given size and experience mix of the aviator force. Second, the change in AvIP could change the overall size of the force and the experience mix of the force by changing retention behavior. Such a change could affect the Army's pay bill and accrual charge for the aviator force, as well as AvIP costs. Our cost computations accounted for both types of changes.

One limitation of our analysis was that, because of computing constraints and the number of assumptions required,[1] we did not include the effort exerted by officers to reach milestones and, therefore, the disutility associated with putting more effort into reaching milestones under alternative pay schemes, which could change our results. In some of our simulations, we changed AvIP in a way that could potentially incentivize individuals to put more effort into achieving milestones, which theoretically would increase the probability of milestone achievement. If so, it is likely that the retention-maintaining values of AvIP that we found, described in the next section, are too high, because more people would reach each milestone. The change in milestone achievement would also affect AvIP cost. However, we did perform sensitivity analyses to determine how the results might change if the probability of reaching different milestones changes.

Simulated Retention Effects of Alternative Values of AvIP

Discussions with the Army revealed two major concerns about AvIP: that the lack of updating the nominal AvIP value for over 20 years had decreased its real value for maintaining an inventory of qualified personnel and that AvIP is the same for everyone regardless of whether they reach aviation career milestones in a timely fashion.

Influence of Inflation on AvIP's Retention Effects

It was shown in Chapter 2 that because AvIP remained unchanged between 1998 and July 2020, the real value of AvIP fell as a result of inflation. To predict how the reduction in the real value of AvIP would have affected retention in the Army Aviation Branch, and how those effects may have been ameliorated by the increase in AvIP in January 2020, we simulated the differences in retention under three scenarios: (1) a baseline scenario in which AvIP was set equal to its real value in 2019, (2) a scenario in which AvIP was set so as to maintain its real value as of 1998,[2] and (3) a scenario in which AvIP was set equal to its real value after the increase in January 2020. In each of the scenarios, it was assumed that individuals served their entire careers with AvIP set equal to that value.

Figure 5.1 shows the simulated retention curves for commissioned officers (left) and for warrant officers (right) in the 2021 entering cohort who faced a ten-year initial ADSO. The model predicts that had AvIP maintained its 1998 real value from FY 1998, commissioned officer retention would have been 1.3 percent higher and warrant officer retention would have been 3.8 percent higher than in the baseline (FY 2019 real value) scenario. Retention when AvIP is set equal to its real value as of January 2020 is just slightly higher than in the baseline: 0.6 percent higher for commissioned officers and 1.6 percent higher for warrant officers. In other words, retention when AvIP is set equal to its real value as of January 2020 is lower than when it is set equal to its real value as of 1998 because the real value of AvIP after the January 2020 increase is still below its real value in 1998.[3]

---

[1] See Asch et al. (2020) for a discussion of the assumptions and computing power involved in simulating effort.

[2] Real value in 1998 was calculated using the CPI-U Inflator; that is, AvIP × CPI-U (2019)/CPI-U (1998).

[3] Under a six-year ADSO, if the AvIP had maintained its real value from FY 1998, then retention would have been 1.9 percent higher for commissioned officers and 5.4 percent higher for warrant officers over their entire careers than it would have been over an entire career under the real values in FY 2019 (the baseline). Under the January 2020 values, retention would be 0.8 percent higher for commissioned officers and 2.3 percent higher for warrant officers over their entire careers. See Figure D.1 in Appendix D.

FIGURE 5.1
Simulated Retention If AvIP Maintained 1998 Real Value in 2019

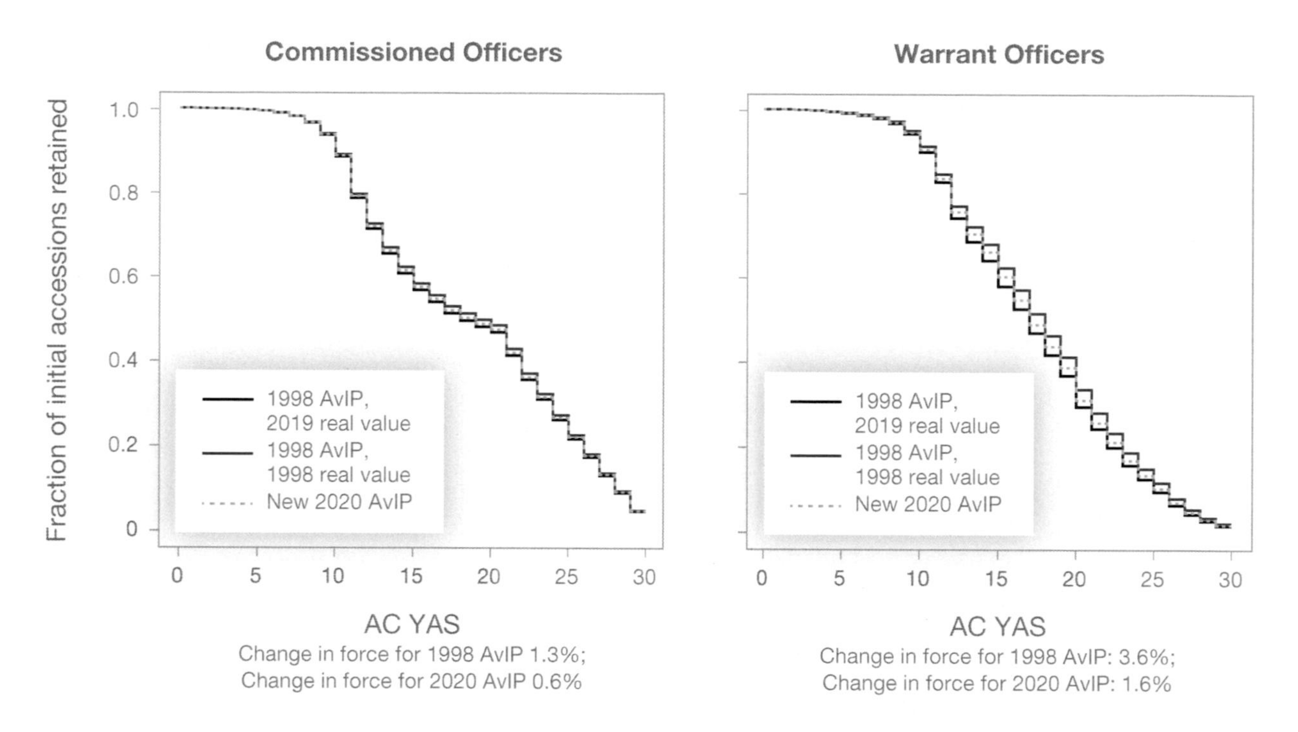

NOTE: Figures report retention of commissioned officers (left) and warrant officers (right) in the Army Aviation Branch under various levels of AvIP. Baseline (black line) refers to the value of AvIP in force from FY 1998 to July 2020 (see Table 2.2). Real value in 1998 (red line) was calculated using the CPI-U Inflator, that is, AvIP * CPI-U (2019)/CPI-U (1998). New 2020 AvIP (blue dashed line) refers to the value of AvIP in force starting January 2020 (see Table 2.3). Simulations assumed a ten-year initial ADSO that tracked warrant officers with 7 to 13 or 19 to 22 years of AFS who are eligible for $30,000 per year in AvB over a three-year contract. See Figure D.1 for analogous results under a six-year initial ADSO.

Milestone-Based AvIP

The Army expressed concern that AvIP does not reward career progression—that is, everyone earns the same value of AvIP, regardless of whether they attain career milestones in a timely manner. The Army requested that we simulate how retention would change if AvIP was based solely on milestones. To assist with this analysis, the Army provided us with prototype AvIP dollar amounts for each proposed milestone, recognizing that those amounts are subject to change depending on personnel or cost requirements. We evaluated how retention would be affected by adoption of these prototype values. We also calculated AvIP levels that maintain retention at its baseline. We also considered how the retention-maintaining values change with heterogeneity in the timing or probability of milestone achievement. Finally, we simulated how the retention-maintaining values might change if it becomes easier to achieve certain aviation milestones.

Values of AvIP by Milestone

The Army's prototype milestone-based values of AvIP are shown in Table 5.1. We present the results of our simulation of the retention effects of the Army's prototype AvIP values in Figure 5.2 for commissioned officer aviators (left) and warrant officer aviators (right), as well as in column 2 of Table 5.2. In this scenario and all following simulations in this chapter, we treated baseline AvIP as being the AvIP values in effect as of FY 2020 (*current AvIP*).

For commissioned officers, the Army's prototype AvIP values reduce retention and reduce the size of the commissioned officer aviator inventory by 1 percent relative to the baseline where AvIP is set to the 2020

TABLE 5.1
The Army's Prototype Milestone-Based AvIP Values

Milestone	Monthly AvIP Value ($)
Flight school entry	125
Pilot status	200
Pilot-in-command	400
Captain's Career Course (CO) or track (WO)	700
Senior aviator	1,000
Master aviator	1,000 (0 for COs with > 25 YAS)

SOURCE: Provided to authors by U.S. Army Aviation Center for Excellence, 2022.
NOTES: CO = commissioned officer; WO = warrant officer.

values. Much of the reduction is concentrated in mid-career, after aviators complete their ten-year ADSO. We estimated that AvIP costs to the Army for commissioned officer aviators would fall by 24.2 percent; therefore, overall aviator personnel cost, including the cost of military pay and the Army's retirement accrual charge, would fall by 0.6 percent.

For warrant officers, the Army's prototype AvIP values reduce retention by 1.9 percent, AvIP cost by 16.3 percent, and overall cost by 0.4 percent.[4] Retention falls because a rising AvIP over the first two decades of a career is no longer guaranteed as it is under the baseline. That is, although the Army's prototype values are comparable to the current baseline AvIP values, not all aviators under the prototype AvIP receive those values because not everyone who reaches a given YAS also achieves a given milestone. In contrast, under the baseline, everyone receives the AvIP amount if they reach a given YAS.

Because the Army's prototype milestone-based AvIP values are predicted to reduce retention, we considered how those values might be modified to maintain retention. In particular, we searched for AvIP values at each milestone that produced the smallest change in overall force size relative to the baseline, which in all cases is smaller than 0.003 percent. Table 5.2 reports those results, with the Army's prototype results reported in column 2 and with all percentage changes reported in the table being relative to the current AvIP values. These retention-maintaining values of AvIP are shown in Table 5.2 for commissioned officers (column 3) and for warrant officers (column 5). The retention-maintaining values are in many cases substantially higher than the values proposed by the Army, suggesting that higher amounts per milestone are needed to offset the fact that not everyone reaches the milestones under a policy that bases AvIP on achievement of milestones. We note that while the overall change in retention is small, there could be small differences in the experience mix that affect the likelihood of reaching a particular YAS (and therefore the likelihood of reaching a particular milestone). For commissioned officers, the retention-maintaining AvIP values decrease force size prior to 20 YAS by −0.02 percent and increase force size after 20 YAS by 0.14 percent. For warrant officers, the retention-maintaining AvIP values decrease force size prior to 20 YAS by −0.07 percent and increase force size after 20 YAS by 1.66 percent. Furthermore, the retention-maintaining values raise costs relative to the current AvIP: AvIP costs rise by 14.57 percent for commissioned officers and 9.31 percent for warrant officers, whereas overall costs rise by 0.38 percent for commissioned officers and 0.20 percent for warrant officers.

[4] Under a six-year ADSO, for commissioned officers, the Army's prototype AvIP values lower retention by 1.5 percent, AvIP cost by 21.8 percent, and overall cost by 0.42 percent. For warrant officers, the Army's prototype AvIP values lower retention by 1.6 percent, AvIP cost by 10.2 percent, and overall cost by 0.17 percent. See Figure D.2 in Appendix D.

FIGURE 5.2
Simulated Retention Effects of the Army's Prototype Milestone-Based AvIP Values

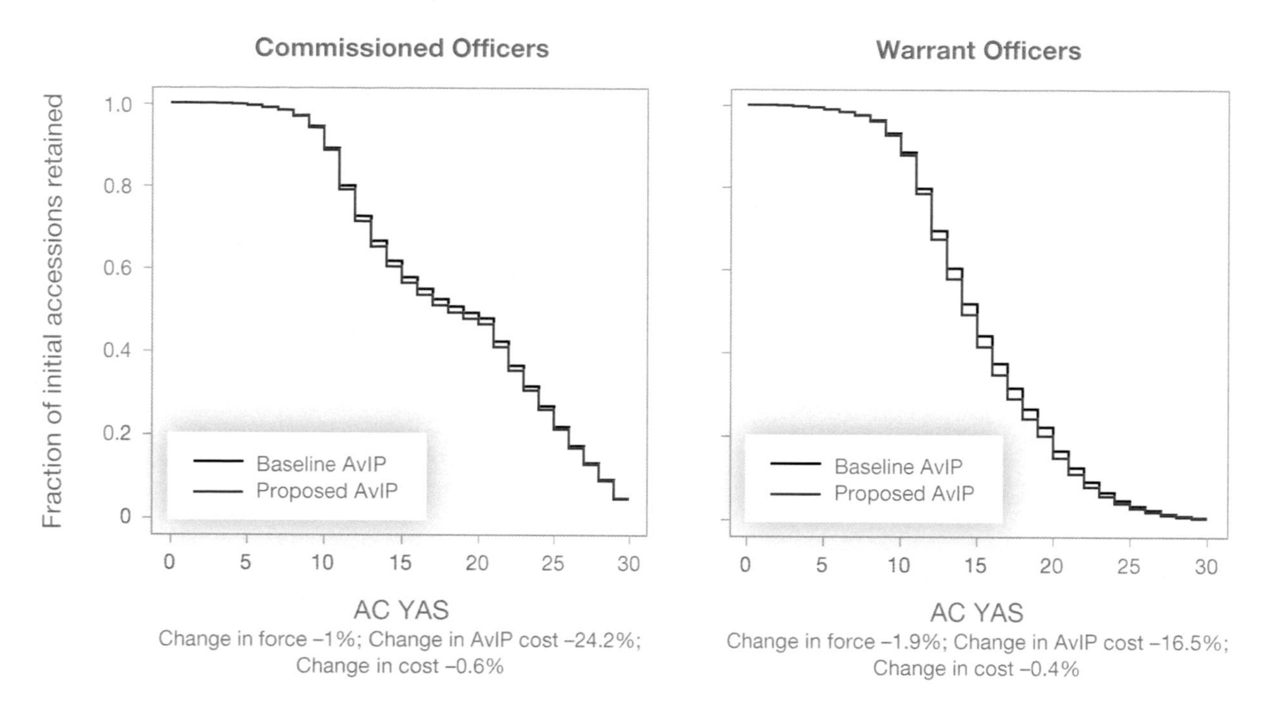

NOTE: This figure shows simulations of retention of commissioned officers and warrant officers in the Army Aviation Branch under different ways of setting AvIP. The black line shows retention under the AvIP values set in January 2020 (see Table 2.3). The red line shows retention under the Army's prototype AvIP values (see Table 5.1). Simulations assume a ten-year initial ADSO and that AvB is not offered to warrant officers. See Figure D.2 for analogous results under a six-year initial ADSO.

Note that, for commissioned officers, the retention-maintaining values of AvIP that we identified would create a perverse incentive by decreasing the reward to achieving more-senior milestones. Because AvIP decreases from senior aviator to master aviator, commissioned officers may attempt to avoid making master aviator. We do not account for this effect in our analysis because our modeling does not include the decision to exert the effort, and the disutility of that effort, to reach each milestone.

Column 4 of Table 5.2 shows the retention effects of removing the perverse incentive. Unlike in column 3, AvIP for master aviator remains at $1,575 rather than declines to $1,460. We note that this alternative will lead to the payment of economic rents by the Army; i.e., the Army will pay an additional amount for achieving master aviator to individuals who would have chosen to make master aviator even without the additional amount. Removing the perverse incentive raises overall force size by 0.02 percent, raises AvIP cost by 16.4 percent, and raises overall cost by 0.43 percent relative to the baseline.

Our results are subject to several caveats. We note that there may be multiple values of AvIP that maintain or nearly maintain retention. Furthermore, the values we find may not be cost-minimizing; the optimization procedure was too complicated to implement without simplifications, such as having AvIP increase by the same dollar amount for each milestone. As previously discussed, our results also do not take into account the effort-exertion decision and therefore the incentive effect of basing AvIP on milestones. In addition, our results depend on assumptions regarding the timing and probability of milestone achievement; our results could change if our estimated probabilities are incorrect (such as because of changes over time or data quality issues), if constraints on officers' ability to achieve milestones change, or if there is heterogeneity in timing

Table 5.2. Retention-Maintaining Values of Milestone-Based AvIP

		Commissioned Officers		
AvIP Value by Milestone	Army Prototype	Retention-Maintaining AvIP	Retention-Maintaining Without Perverse Incentive	Warrant Officers
Flight school entry	$125	$125	$125	$125
Pilot status	$200	$300	$300	$200
Pilot-in-command	$400	$600	$600	$435
Captain's Career Course (CO) or Track (WO)	$700	$1,040	$1,040	$740
Senior aviator	$1,000	$1,575	$1,575	$1,600
Master aviator	$1,000[a]	$1,460[a]	$1,575[a]	$1,690
% change in force size relative to current AvIP	–1.0% (COs), –1.9% (WOs)	0.002%	0.024%	0.000%
% change in force size before 20 YAS relative to current AvIP	–0.8% (COs), –1.4% (WOs)	–0.020%	–0.009%	–0.065%
% change in force size after 20 YAS relative to current AvIP	–2.8% (COs), –13.7% (WOs)	0.142%	0.233%	1.658%
% change in AvIP cost relative to current AvIP	–24.19% (COs), –16.46% (WOs)	14.570%	16.400%	9.310%
% change in overall cost relative to current AvIP	–0.63% (COs), –0.41% (WOs)	0.380%	0.430%	0.200%

NOTE: This table reports results of simulations of retention of commissioned officers in the Army Aviation Branch under different ways of setting AvIP, with AvIP values for each milestone reported in the columns. Baseline retention uses the AvIP values set in January 2020 (see Table 2.3). Simulations assume a ten-year initial ADSO and that AvB is not offered to warrant officers. See Table D.1 for analogous results under a six-year initial ADSO. CO = commissioned officer; WO = warrant officer.

[a] AvIP is $0 for commissioned officers with over 25 YAS.

or probability of milestone achievement. We therefore interpret the results presented in this section as an approximation only, and we provide several sensitivity checks in Appendixes D and F.[5]

Excursion: What If Milestone Achievement Became More Common?

In this sub-section we consider the effect of higher probabilities of achieving milestones. The probability of achieving milestones changes over time. For instance, discussions with our sponsor revealed that, in the early 2000s, commissioned officers were not required to make pilot-in-command. The measured probabilities and timing are also subject to changes in constraints on officers' ability to achieve milestones (e.g., changes in the

[5] Under a six-year initial ADSO, both commissioned and warrant officers have lower retention-maintaining AvIP values at pilot status, pilot-in-command, and Captain's Career Course or track. Commissioned officers have slightly higher values at senior and master aviator, whereas warrant officers have slightly lower values. Both AvIP and overall cost also increase by less when the simulation is run under a six-year initial ADSO. See Table D.1. Under heterogenous milestone timing, the retention-maintaining AvIP values tend to decrease slightly (typically by under $25 per month for each milestone), and the changes in AvIP and overall cost are similar; see Table F.1. Under heterogenous milestone achievement probability, retention-maintaining AvIP decreases at every milestone for warrant officers (by up to $90 per month for some milestones) and decreases at pilot status, pilot-in-command, and master aviator but increases slightly at Captain's Career Course and substantially at senior aviator for commissioned officers. AvIP cost also increases by more (16.78 percent) for commissioned officers and less (6.32 percent) for warrant officers; see Table F.2.

difficulty of getting the necessary flight hours to achieve pilot-in-command) and to data quality issues (e.g., as noted in Chapter 2 tracking is denoted in TAPDB using an SQI, but we see only the duty SQI for warrant officers rather than whether an individual has attained a particular SQI, which may lead to the undercounting of tracked warrant officers).

To show that raising the probability of milestone achievement changes the retention-maintaining AvIP values, we ran a simulation where we raised the probability of achieving pilot-in-command for commissioned officers to 90 percent in both baseline and the simulation instead of 65 percent as shown in Table 2.1 (assuming that all commissioned officers share the same timing and probability of milestones). Similarly, we continued to assume that all warrant officers shared the same timing and probability of milestones, but we raised the probability of achieving track to 90 percent in both baseline and the simulation instead of 68 percent as shown in Table 2.1. These simulations could correspond to real-world scenarios where, for instance, there are more opportunities for commissioned officers to earn flight hours and where there are more seats in any training classes needed for warrant officers to track.

The retention-maintaining AvIP values under these new probabilities are reported in Table 5.3. We found that if it becomes easier for commissioned officers to reach pilot-in-command, then the retention-maintaining values of AvIP drop substantially, the change in AvIP cost is only 8.1 percent compared with 14.6 percent as shown in Table 5.2, and the change in overall aviator personnel cost is only 0.21 percent compared with 0.38 percent. Similarly, if it becomes easier for warrant officers to reach tracking, then the retention-maintaining values of AvIP drop substantially for several milestones, the change in AvIP cost is only 4.6 percent compared with 9.3 percent as shown in Table 5.2, and the change in overall aviator personnel cost is only 0.1 percent compared with 0.2 percent.[6] The reason why the retention-maintaining values are lower is that more officers receive AvIP when the likelihood of achieving the milestone is higher. Consequently, the dollar value can be lower to sustain overall retention.

We found that the values of milestone-based AvIP would not need to increase as much to maintain retention if the probability that Army aviators achieved career milestones are higher than observed in the data. Thus, to the extent that basing AvIP on milestones increases the likelihood that a pilot achieves the milestone, the increase in cost and the degree to which AvIP must be higher is reduced.[7] Note that human capital achievement could increase either because of larger investments on the part of the aviator or because of policy changes by the Army; our simulations do not distinguish between the two, but an analysis of how basing S&I pays on human capital achievement would affect human capital investment could be an interesting avenue for future work.

Potential for AvIP to Retain High-Ability Personnel

One impetus for considering AvIP using attainment of career milestones is that an appropriately structured incentive pay might induce higher-ability personnel to stay and seek the achievement of each career milestone. In this context, by *ability* we mean characteristics of individual aviators that would tend to increase or decrease their likelihood of achieving different milestones relative to their peers. Such characteristics include innate cognitive intelligence and other characteristics that lead to success in achieving milestones, such as an

[6] This figure does not include any costs associated with raising the probabilities of milestone achievement. If, for instance, raising the probabilities requires additional training seats in some programs, then there may be an associated cost.

[7] Under a six-year initial ADSO, the retention-maintaining AvIP value for pilot-in-command for commissioned officers is $100 per month lower, and the values for tracking, senior aviator, and master aviator for warrant officers are $45–55 per month lower. AvIP cost increases by only 6.76 percent for commissioned officers and 1.97 percent for warrant officers. See Table D.2.

TABLE 5.3

Retention-Maintaining Values of AvIP with Higher Probability of Milestone Achievement

AvIP Value by Milestone	Commissioned Officers	Warrant Officers
Flight school entry	$125	$125
Pilot status	$150	$175
Pilot-in-command	$450	$225
Captain's Career Course or track	$810	$720
Senior aviator	$1,200	$1,340
Master aviator	$1,200[a]	$1,375
% change in overall force size relative to current AvIP	–0.002%	–0.001%
% change in force size before 20 YAS relative to current AvIP	–0.032%	0.005%
% change in force size after 20 YAS relative to current AvIP	0.189%	–0.171%
% change in AvIP cost relative to current AvIP	8.090%	4.560%
% change in overall cost relative to current AvIP	0.210%	0.100%

NOTE: Values reflect a 90-percent probability of achieving pilot-in-command (commissioned officers) or track (warrant officers). This table reports results of simulations of retention of commissioned officers in the Army Aviation Branch under different ways of setting AvIP, with AvIP values for each milestone reported in the columns. Baseline retention uses the AvIP values set in January 2020 (see Table 2.3). Simulations assume a ten-year initial ADSO and that AvB is not offered to warrant officers. See Table D.2 for analogous results under six-year initial ADSO.

[a] AvIP is $0 for commissioned officers with over 25 YAS.

ability to work well in the aviator community. We recognize that achievement of a milestone (or, conversely, the lack of achievement of a milestone) might be because of random factors outside an individual's innate ability, and this is reflected in our modeling approach where milestone achievement is stochastic. To explore how setting AvIP using career milestone achievement could affect the retention of higher-ability aviators, we extended the DRM following an approach similar to the approach used in Asch, Mattock, and Tong (2020) and Asch et al. (2021). We drew heavily from these reports in the following exposition.

Asch and Warner were the first to incorporate ability into a DRM, and they used the model to assess the retention, performance, and cost effects of alternative retirement reform proposals, as well as policies to restructure the military pay table (Asch and Warner, 1994, 2001). In particular, in their model, higher-ability personnel are promoted faster and have higher promotion probabilities, but higher-ability personnel also have better external opportunities. Compensation policy can affect the financial benefits of staying for higher-ability personnel. Asch and Warner used their DRM to provide simulations of how compensation reforms affected overall retention, the retention of higher-ability personnel, ability sorting into higher grades, and personnel cost.

The Asch-Warner simulations were based on a calibrated model the key parameters of which, such as the mean and standard deviation of taste for service, were assumed to replicate the observed retention profile. In contrast, the parameters of the DRM used in this project were estimated, not calibrated. We built on the Asch and Warner modeling of ability and incorporated their approach into our DRM simulation capability to evaluate alternative AvIP policies.

In a departure from the Asch-Warner approach, we considered milestone attainment speed, rather than speed of promotion, as being related to ability. This was done to capture the effect of an AvIP policy that would make the value of AvIP dependent on the latest career milestone that a service member has achieved rather than being dependent on the YAS.

Adding Ability to the DRM Simulation Capability

We were able to use the structure of the DRM, along with the estimated parameters and assumptions about how innate ability affects milestone attainment speed, to examine how selectively alternative AvIP structures affect ability. To incorporate ability into the DRM, we assumed[8]

- the extent to which ability differs among aviator entrants[9]
- the extent to which ability affects the speed with which individuals attain career milestones
- the effect of ability on external civilian opportunities.

We discuss each of these in turn.

First, we assumed that any given individual has a fixed level of ability at entry, drawn from a normal distribution. The standard deviation of the distribution indicates the extent to which ability differs among military entrants. The values of the mean and standard deviation for the distribution we used in our simulations were calibrated to replicate the steady-state retention profiles of commissioned officers and warrant officers, given the other two assumptions we made.

Second, we assumed that higher-ability personnel achieve milestones faster.[10] We implemented this concept by subtracting the rounded product of the draw from the normal distribution for a given individual multiplied by one-half from the time-in-service between each milestone. This increase in milestone attainment speed is modeled to start happening between achieving pilot status and pilot-in-command. Thus, a warrant officer with an innate ability of 2 would be one year faster than average to pilot-in-command, two years faster to track, and so on.[11] Consequently, the effect of ability on the speed to the later milestones is larger than for the earlier milestones because the effects on milestone achievement timing are cumulative. Note that we were limited to integer values for the increase or decrease in milestone achievement time because of the time resolution of our model being one year; thus, we could not model achievement being 0.5 years faster, or 1.5 years slower, for example.[12] Figure 5.3 shows how years to milestones vary with ability for Army warrant officers in

---

[8] In Chapter 4 of Asch, Mattock, and Tong (2020), sensitivity analysis was conducted to determine how varying the assumptions about how ability enters the model affected the simulations considered. The study found that varying the assumptions did not affect the qualitative results.

[9] We assumed the distribution of ability at entry to be fixed and the same under alternative AvIP structures.

[10] We also conducted exploratory simulations where we assumed that higher-ability personnel had a greater probability of achieving milestones and had qualitatively similar results to those reported below.

[11] The need to round to the closest year can lead to some unevenness in the effect of ability on milestone attainment speed. For example, a warrant officer with an innate ability of 1 would not be any faster than average to pilot-in-command (as round($0.5*1$) = 0, although this is computer implementation-dependent), but they would be one year faster to track (as round($1*1$) = 1), two years faster to senior aviator (as round($1.5*1$) = 2, although this too is computer implementation-dependent), and finally two years faster to master aviator (as round($2*1$) = 2).

[12] This limited the sensitivity analysis we could do, because we could not vary the speed by increments smaller than a year. In related work, we examined the effect of reducing the sensitivity of promotion time to ability in a recent paper for the 13th Quadrennial Review of Military Compensation (QRMC) on a time-in-grade pay table (Asch, Mattock, and Tong, 2020) by having ability affect time to promotion to E-7 and above (rather than E-6 and above) and found that the results remained qualitatively the same in our comparison of average ability percentiles by grade across the time-in-grade and time-in-service pay tables.

FIGURE 5.3
Years to Career Milestones, by Ability Level, for Army Warrant Officer Aviators

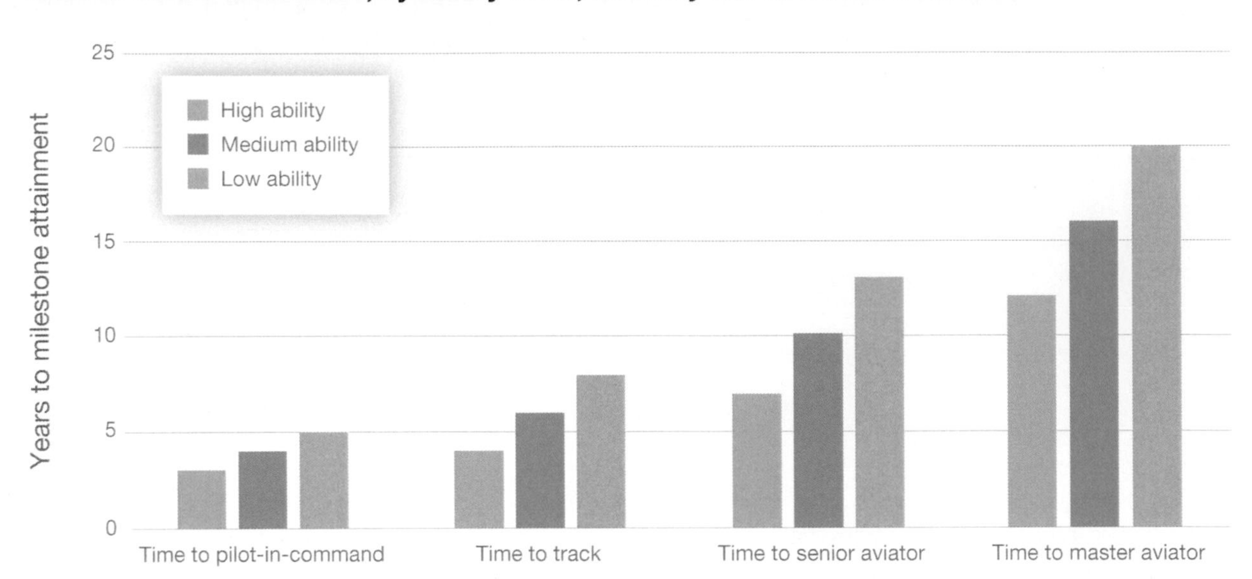

our implementation; the figure for commissioned officers is qualitatively similar. As mentioned in the previous paragraph, the assumed parameters are calibrated to best fit the retention profile for commissioned and warrant officer aviators.[13]

Third, we assumed that higher-ability service members also have better external opportunities. We modeled this by multiplying the civilian opportunity wage by one plus 0.1 (that is, 10 percent, a parameter value that has worked well in past studies [e.g., Asch, Mattock, and Tong, 2020]) times the ability distribution standard deviation times the individual's ability draw, or $(1 + (0.1 \times \sigma_a \times a))$, where σ_a is the standard deviation of the draw, and a is the individual's ability draw. This has the effect of increasing the civilian opportunity wage for high-ability individuals and decreasing the civilian opportunity wage for low-ability individuals. For example, an individual with an innate ability of 1 drawn from a normal distribution with a mean of 0 and a standard deviation of 1.0 would have an opportunity wage that is 10 percent greater than that of the average individual, while an individual with an innate ability of –1 would face a civilian opportunity wage that is 10 percent less.[14] Similarly, if the parameter value was 0.05 (that is, 5 percent), then an individual

---

[13] In effect, this assumption creates different paths for service members with lower ability, medium ability, and high ability. A service member one year slower to one milestone is also promoted more slowly to the following milestones (provided they attain those milestones), and, conversely, a service member one year faster to one milestone is also faster to the succeeding milestones (if they attain them). Never would a service member be faster to one milestone and then slower to the next. The service member knows which path they are on with certainty, be it slow, medium, or fast; the time to the next potential milestone is non-stochastic, with the only residual uncertainty being whether they will attain the milestone or not. This strong assumption made the model tractable, but it may assume away some uncertainty that is important. We hope to have the opportunity to relax this assumption in future work.

[14] Similar to the impact of ability on speed of milestone attainment, impact of ability on the civilian opportunity wage is non-stochastic, though it does depend on the standard deviation of the ability distribution. This strong assumption made the model tractable but potentially assumes away some important uncertainty. We hope to relax this assumption in future work.

While we did not explore sensitivity of this model to the civilian wage assumption, we did explore this sensitivity for a closely related model in our 13th QRMC report on a time-in-grade pay table (Asch, Mattock, and Tong, 2020) and found that the results remained qualitatively the same in our comparison of average ability percentiles by grade across the time-in-grade and time-in-service pay tables.

with an innate ability of 1 would have a civilian opportunity wage 5 percent greater than that of the average individual.

We illustrate how we calibrated the mean and standard deviation of the normal distribution to fit the observed retention profile in Figure 5.4 for Army warrant officer aviators. In the process of calibration, we systematically varied the mean and standard deviation within the DRM and chose the mean and standard deviation that most closely replicated the historically observed retention, as indicated by the Kaplan-Meier curve. The left panel shows the observed retention profile versus the simulated retention profile when we miscalibrated the mean and standard deviation to equal 0 and 3, respectively. The simulated retention profile is too low relative to the observed profile through mid-career and then too high thereafter. We chose a standard deviation of 1 instead, resulting in a good fit, as shown in the right panel. When ability is drawn from a normal distribution with a mean of 0 and a standard deviation of 1, 31 percent of simulated individuals are assigned an ability greater than 0.5 and 16 percent an ability greater than 1 (high ability), while 31 percent of simulated individuals are assigned an ability less than –0.5 and 16 percent an ability less than –1 (low ability). As a result, about two-thirds of the simulated individuals are within 1 standard deviation of the average (or average ability), as would be the case with any instance of the normal distribution.

Ability Retention Simulation Results

As we mentioned above, one impetus for considering a career milestone-based AvIP is that it might be structured to induce higher-ability personnel to stay and lower-ability personnel to leave. At a minimum, we

FIGURE 5.4
Calibrating the Parameters of the Ability Distribution for Army Warrant Officer Aviators

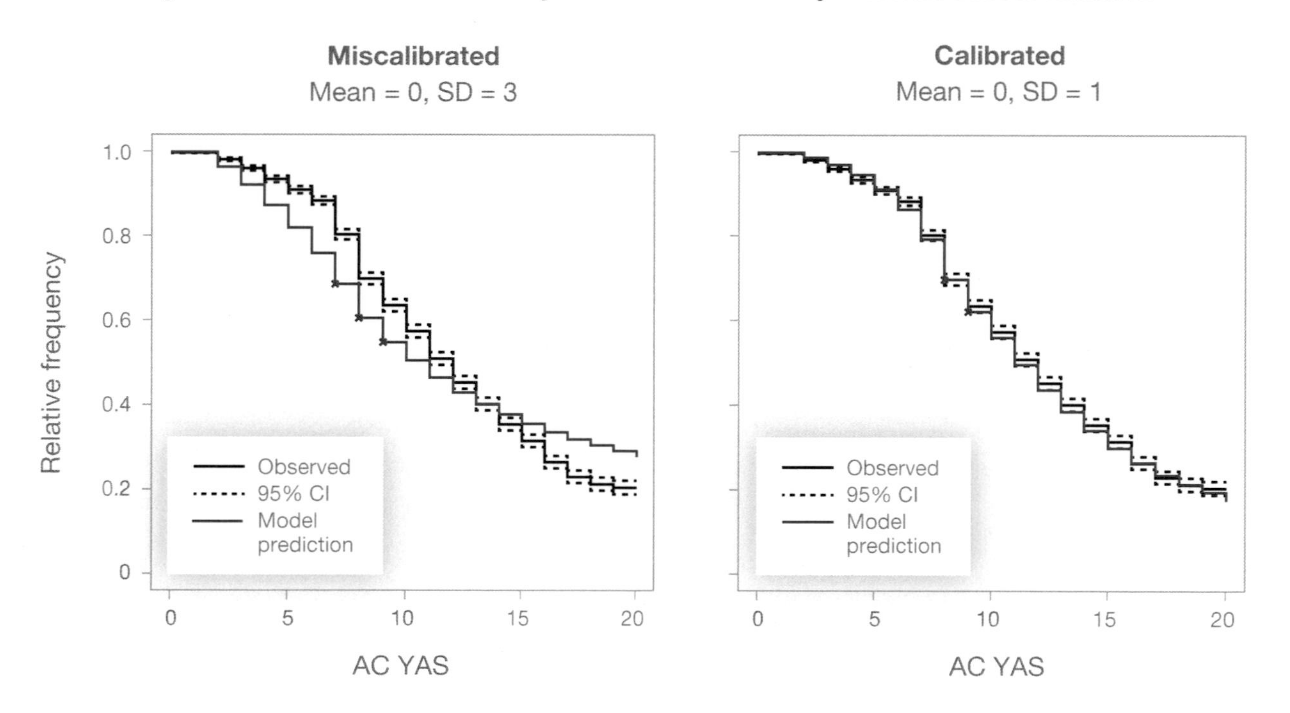

NOTE: This figure displays results for observed and simulated cohorts from 2002–2010. AC = active component; CI = confidence interval; SD = standard deviation.

would want any new AvIP policy to perform at least as well as the existing policy in retaining high-ability individuals and ideally would want it to perform better.

We used the DRM to simulate the effect of alternative AvIP policies on the retention of higher-ability individuals—specifically, the AvIP schedules shown in Tables 5.4 and 5.5. These policies include the baseline and the prototype AvIP values, as well as the retention-maintaining AvIP and some additional alternatives. To measure this effect, we computed the average ability percentile under the baseline policy and under the alternative. Given our assumption of a normal distribution of ability at entry with a mean of 0, the mean ability percentile at entry would be the 50th percentile of the distribution. We then computed the average ability percentile across the force that is retained, as well as the average ability percentile at each milestone. We also looked to see whether there was evidence of improved ability sorting; that is, whether there was an increased tendency for higher-ability individuals to stay and attain higher milestones.

Our main result was that the impact of the AvIP policies considered was small and often mixed, generally affecting the mean percentile retained at each milestone by less than one-half a percentile point. We explored the possible reasons for this lack of impact of an AvIP based solely on milestone attainment and examined the impact of AvIP policies where the AvIP amount is based not only on what milestone is attained but also how quickly it is attained relative to one's peers.

Table 5.4 shows the alternative AvIP schedules. We considered five alternatives to the current baseline AvIP schedule: (1) the Army prototype, (2) the retention-maintaining alternative, (3) a highly-skewed AvIP schedule, and (4) two AvIP schedules that depend on milestone attainment speed, one derived from the retention-maintaining schedule, and (5) one from the skewed schedule. The baseline, prototype, and retention-maintaining AvIP policies have already been discussed. In the skewed AvIP schedule, S&I pay increases at an increasing rate—here, doubles—with each successive milestone, providing a higher incentive for attainment of each successive milestone.

The final two AvIP schedules reward the speed with which milestones are achieved in addition to the milestone itself. Starting with pilot-in-command, service members who are faster than average at attaining a milestone receive a higher AvIP than their slower peers, with the value ranging from 150 percent higher (for the variant derived from the retention-maintaining AvIP) to 185 percent higher (for the variant derived from the skewed schedule). This can be seen in the last two rows of Table 5.4; in the cells where there are two entries, the first is the AvIP value for those who attain the milestone faster, while the second is the value for those who attain the milestones more slowly.

TABLE 5.4
Warrant Officer Milestone-Based AvIP Schedules

Alternative AvIP Schedule	Flight School Entry ($)	Pilot Status ($)	Pilot-in-Command ($)	Track ($)	Senior Aviator ($)	Master Aviator ($)
Prototype	125	200	400	700	1,000	1,000
Retention maintaining	125	200	435	740	1,600	1,690
Skewed AvIP by milestone	83	167	334	668	1,336	2,672
Higher AvIP for faster achievers	122	195	509 339	866 577	1,872 1,248	1,977 1,318
Higher skewed AvIP for faster achievers	79	158	411 221	822 442	1,643 885	3,286 1,770

NOTES: This table presents the monthly AvIP pay schedules used in our simulations. In cells containing two numbers, the first number indicates the AvIP for service members faster than average to that milestone, and the second number gives the AvIP for service members slower than average to that milestone.

Table 5.5 shows the change in the mean ability percentile for the overall population of Army warrant officer aviators, as well as by greatest milestone attained relative to the current baseline AvIP policy shown in Table 2.3. The first row of the table shows the mean ability percentile under the baseline, the second row of the table shows the mean ability percentile under the Army prototype, and the third row shows the difference between the prototype and the baseline. The remaining rows of the table all show differences relative to the baseline. The first thing to note is that the difference from the baseline reported in the "overall" column is positive in every case. For warrant officers, all the alternatives we examined result in the retention of a higher-ability force. The second thing to note is while the overall differences are all positive, they are also relatively modest, amounting to only 1 percentile point or less.

While the differences are all positive in the third row, it is important to remember that retention is not maintained under the prototype. Moving on to the fourth row, where we see the difference between the retention-maintaining AvIP and the baseline, we see that there is a small positive effect on the force overall, but we see also negative numbers for the two highest milestones, indicating that this AvIP schedule has a negative impact on ability sorting. We also see that the impact is small, as all numbers are less than one-half a percentile point different from the baseline.

The fifth row of Table 5.5 shows the results for the skewed AvIP. It shows a higher overall difference than the preceding row (0.4 percentile points compared with 0.2 percentile points), so this schedule proves better at retaining high-ability service members. The differences by milestone are mixed, with some positive and some negative. However, the value for master aviator is 0.7 percentile points. This points to some ability sorting, because the difference associated with pilot status was –0.1.

The sixth row of Table 5.5 shows the results for a schedule derived from the retention-maintaining schedule where individuals who are faster than average to a milestone receive a higher pay. Faster service members receive an AvIP equal to 150 percent of the value received by their slower peers. The overall mean percentile is 0.7 percentile points higher than the baseline, and each successive milestone beyond pilot status is associated with a higher difference from the baseline, ranging from 0.3 for pilot-in-command to 2.2 for master aviator.

TABLE 5.5

Warrant Officer Change in Mean Ability Percentile

Alternative AvIP Schedule	Overall	Flight School Entry	Pilot Status	Pilot-in-Command	Track	Senior Aviator	Master Aviator
Baseline case	49.2	50.3	44.2	47.0	49.8	56.1	63.9
Prototype	49.4	50.4	44.2	47.1	49.9	56.2	64.0
Prototype minus baseline	0.2	0.1	0.0	0.1	0.1	0.1	0.1
Retention maintaining minus baseline	0.2	0.1	0.0	0.1	0.1	–0.4	–0.4
Skewed AvIP by milestone minus baseline	0.4	0.1	–0.1	0.1	0.1	–0.2	0.7
Higher AvIP for faster achievers minus baseline	0.7	0.1	0.0	0.3	0.4	0.9	2.2
Higher skewed AvIP for faster achievers minus baseline	1.0	0.1	–0.1	0.3	0.6	1.7	4.6

NOTES: Ability is a unitless measure in the model with an assumed mean and standard deviation for the accession cohort. We computed the percentile of the ability distribution for each service member in the force.

The final row of Table 5.5 shows results for the skewed schedule where the level of AvIP depends on speed of milestone attainment. In this case, faster service members receive an AvIP equal to approximately 185 percent of the value received by their slower peers. The overall ability mean is 1.0 percentile point more than the baseline, and we see a strong ability-sorting effect with the mean difference ranging from −0.1 percentile point at pilot status to 4.6 percentile points for the master aviator milestone.

Tables 5.6 and 5.7 show our results for commissioned officers; Table 5.6 shows the milestone-based AvIP schedules and Table 5.7 shows the changes in mean ability percentiles. The results tend to be similar to those we saw with warrant officers, but somewhat diminished. This may, in part, be because commissioned officers are less responsive to this S&I pay, which constitutes a smaller fraction of their overall earnings. There are some differences though, the most striking of which is that the alternative AvIP schedules do not uniformly

TABLE 5.6

Commissioned Officer Milestone-Based AvIP Schedules

Alternative AvIP Schedule	Flight School Entry ($)	Pilot Status ($)	Pilot-in-Command ($)	Track ($)	Senior Aviator ($)	Master Aviator ($)
Prototype	125	200	400	700	1,000	1,000
Retention maintaining	125	300	600	1,040	1,575	1,460
Skewed AvIP by milestone	110	219	438	876	1,752	3,504
Higher AvIP for faster achievers	129	311	745 497	1,292 861	1,956 1,304	1,813 1,209
Higher skewed AvIP for faster achievers	106	212	551 297	1,102 594	2,205 1,187	4,410 2,374

NOTES: This table presents the monthly AvIP pay schedules used in simulations. In cells containing two numbers the first number indicates the AvIP for service members faster than average to that milestone, and the second number gives the AvIP for service members slower than average to that milestone.

TABLE 5.7

Commissioned Officer Change in Mean Ability Percentile

Alternative AvIP Schedule	Overall	Flight School Entry	Pilot Status	Pilot-in-Command	Captain's Career Course	Senior Aviator	Master Aviator
Baseline case	50.9	50.5	49.4	40.7	52.2	55.3	62.0
Prototype	50.8	50.4	49.4	40.7	52.1	55.3	61.9
Prototype minus baseline	−0.1	−0.1	0.0	0.0	−0.1	0.0	−0.1
Retention maintaining minus baseline	−0.1	−0.1	−0.1	0.0	−0.1	−0.1	0.0
Skewed AvIP by milestone minus baseline	0.0	−0.1	0.0	0.0	0.0	0.0	−0.1
Higher AvIP for faster achievers minus baseline	0.1	−0.1	−0.1	0.0	0.2	0.3	0.3
Higher skewed AvIP for faster achievers minus baseline	0.2	−0.1	0.0	0.0	0.3	0.6	1.3

NOTES: Ability is a unitless measure in the model with an assumed mean and standard deviation for the accession cohort. We computed the percentile of the ability distribution for each service member in the force.

improve overall ability. We only saw an improvement in overall ability under the speed-of-attainment–based AvIP schedules. We also only saw positive ability sorting under these two schedules for faster achievers.

Why do the first three alternatives of the milestone-based AvIP not result in more clear-cut ability-sorting results? Some possible reasons include the long initial service obligation associated with flight school, the low probability of attainment for certain milestones, the monetary value of AvIP not being high enough to provide an incentive effect for higher-ability individuals, and individual preferences dominating the pecuniary effect of this S&I pay. In sensitivity analysis not shown here, we found similar results under a six-year ADSO. Similarly, increasing the probability of attainment of a milestone showed little effect. The final two hypotheses are still in contention, and they are both consistent with the levels of AvIP under consideration here being too small to elicit the type of ability-sorting effects we have seen in other work (Asch, Mattock, and Tong, 2020; Asch et al., 2021). The key factor that distinguishes the schedules that result in an ability-sorting effect is that the faster attainment of a milestone results in an enduring advantage in compensation; for the other AvIP schedules, the advantage of faster achievement lasts only as long as it takes for the remainder of a service member's cohort to catch up.[15]

Summary of Findings on AvIP

This chapter provided information on the simulated retention effects of various alternative S&I pays for Army aviators. We first simulated the retention effects of the fall in the real value of AvIP from 1998 to 2020, then the retention effects of switching to a version of AvIP which depends solely on reaching aviation career milestones. We then examined whether switching to having AvIP depend only on milestones would allow the Army to induce higher-quality personnel to stay. The key findings from our simulations of alternative S&I pays are as follows:

- *Setting AvIP based solely on milestones would require higher values than are currently offered at the corresponding YAS values if the Army wishes to maintain retention and will also require setting AvIP differently for commissioned officers and warrant officers.* For instance, as noted in Chapter 2, only 56 percent of commissioned officers who remain in the Army until 10 YAS and who have achieved all previous milestones make senior aviator. As a result, under a policy where aviators receive AvIP solely using milestones, and where the value received roughly corresponds to the current AvIP values for the YAS at which each milestone is achieved, a substantial number of commissioned officers will receive lower AvIP than they do under the baseline policy. We would therefore expect retention of the officers who do not achieve the milestone to decrease under this policy. However, because officers who *do* achieve the milestone are not receiving higher AvIP under the Army's prototype than they do under the baseline, overall retention falls. If the Army wishes to maintain retention, then the AvIP values for each milestone will generally need to be higher than the AvIP values proposed by the Army. This would require a policy change, because AvIP is currently capped at $1,000 per month. We also found different retention-maintaining levels for warrant officers and commissioned officers, suggesting that the Army should consider setting separate values for warrant officers and commissioned officers if it wishes to maintain retention at its current level.

[15] This is similar to the difference in the ability-sorting effect seen in simulations of a time-in-grade pay table versus a time-in-service pay table, where ability is associated with speed of promotion rather than milestone attainment—individuals promoted early in a time-in-grade pay table see an enduring advantage in compensation, which leads to better ability sorting (Asch, Mattock, and Tong, 2020).

- *The retention-maintaining values of milestone-based AvIP will raise AvIP costs relative to their current level.* The Army's prototype AvIP reduces the cost of AvIP by reducing retention. Because the AvIP values for the various retention-maintaining schemes are higher than the Army's prototype, or the current values for each YAS, maintaining retention with AvIP will cost more than the current AvIP plan.
- *If the likelihood of achieving milestones rises, then the Army will maintain retention with lower values of AvIP for each milestone and at a lower cost than if the likelihood of achieving milestones does not change.* We estimated that if commissioned officers were more likely to achieve pilot-in-command and if warrant officers were more likely to track, then the retention-maintaining AvIP values for each milestone would be lower and AvIP costs would rise less relative to costs under the current AvIP values. Thus, to the extent that basing AvIP on milestones increases the likelihood that a pilot achieves the milestone, the increase in cost and the degree to which AvIP must be higher will be reduced relative to the increase in cost and the level of retention-maintaining AvIP under the historical likelihood of milestone achievement.
- *Setting AvIP based on career milestones can increase the incentives for higher-ability pilots to remain in service.* To the extent that higher-ability aviators achieve milestones faster, basing AvIP on milestones also increases the incentive for higher-ability aviators to stay in service. We considered two additional proposals that base AvIP on career milestones. Both have the feature that higher-ability personnel who achieve milestones sooner receive more AvIP sooner than lower-ability personnel. Under the first variant, the value of AvIP values increase in a way that is proportional to the retention-maintaining AvIP schedule. Under the second, AvIP values double at each successive milestone. Under these proposals that increase the reward for achieving milestones, especially for higher-ability aviators, we found that the retention of higher-ability personnel increases, and we found ability sorting improves. Even more higher-ability personnel are retained at later milestones under these proposals than under the baseline current system.

In sum, our analysis suggests that setting AvIP based on career milestones could sustain Army aviator retention, but at a higher cost to the Army than the current values of AvIP. Maintaining retention under a milestone-based AvIP would also require setting some AvIP values higher than the current legal cap of $1,000 per month. We found that increasing the probability of milestone achievement would increase costs by less. Finally, we found that basing AvIP on milestones could increase ability sorting but doing so would require that AvIP sufficiently rewards the earlier achievement of milestones by higher-ability personnel.

CHAPTER 6

Simulated Retention Effects of Alternative Values of AvB

AvB is a flexible tool that Army Aviation can use to prevent shortfalls in force size, whether those are because of previous policies or outside circumstances, such as economic changes, which are known to reduce retention. It is helpful for the Army to know what change in retention can be expected from setting AvB at different levels especially given the civilian demand for pilots described in Chapter 3. The Army requested that we conduct simulations of alternative values of S&I pays by simulating the retention effects of different levels of AvB. This chapter provides the results of that analysis, as well as an analysis of how changes in civilian opportunities affect warrant officer retention when AvB is and is not offered.

We note that, as with the simulations in Chapter 5, simulations in this chapter were run using an assumption of a ten-year initial ADSO. However, the model was estimated using a six-year initial ADSO, and our simulations assumed the same distribution of tastes for active service under both ADSO lengths. Because the change in ADSO length may have changed the distribution of tastes for active service, this is a key limitation in our model. For this reason, Appendix D provides results for the same simulations presented here under the six-year initial ADSO.

Simulated Retention Under Different Values of AvB

As noted in Appendix B, in recent years the value of AvB has differed by aviator milestone; those who have never reached pilot-in-command are ineligible, tracked warrant officers are often eligible for higher amounts than untracked warrant officers, and in some years, untracked warrant officers are not eligible for AvB at all. We simulated the change in predicted force size, relative to a baseline of no AvB, if tracked warrant officers were allowed each level of AvB in multiples of $5,000 per year over a three-year contract, up to the highest value of $50,000 per year. At the request of our sponsor's office, we did not allow untracked pilots-in-command to receive AvB, because no aviators are eligible for AvB prior to the completion of their ten-year initial ADSO, and aviators who have not tracked by 10 YAS are unlikely to ever track.[1] In each case, warrant officers who are no longer under an initial ADSO, have less than 14 or between 19 and 22 years of AFS, and have reached the specified milestone are offered AvB on a three-year contract with the specified yearly bonus. The predicted change in force size can be interpreted as the causal effect of offering AvB at the simulated level on aviator retention, relative to a baseline where no AvB is offered, holding all other factors affecting aviator retention constant.

[1] Since the change to a ten-year ADSO for the 2021 cohort, AvB has not been offered to untracked pilots-in-command (though we note that warrant officers who had 6 YAS in FY 2021 were on the verge of completing their initial ADSO, because only new entrants were subject to the change in ADSO).

The results of our simulations of alternative values of AvB are reported in Table 6.1, where the baseline is no AvB offered. Perhaps as expected, force size rises as yearly AvB rises. As shown in Appendix D, the retention effects of AvB are higher under a six-year initial ADSO, likely because warrant officers who were subject to a six-year initial ADSO have more years of eligibility for AvB. In recent years, the Army has offered approximately $30,000 in AvB for tracked warrant officers (though there is variation by MOS and aircraft type; see Appendix B), which, based on the results shown in Table 6.1, should produce a 6.2 percent increase in overall force size over not offering AvB.

The Effect of AvB on Retention in the Face of Economic Changes

One of the key advantages of AvB to the Army is that it can be flexibly used to maintain retention in the face of economic changes or other factors that might lower retention. Therefore, we also examined how changes in the unemployment rate and civilian pay (relative to military pay) affect retention, both with and without AvB. We typically expected that a falling unemployment rate would lower retention, as would a rise in civilian pay relative to military pay because both represent an improvement in the civilian opportunities available to pilots relative to staying in the Army.

Table 6.2 reports the simulation results when the unemployment rate is between 1 and 10 percent, with the change in force size reported relative to a baseline where the unemployment rate is set to its level in 2018 (4.1 percent; see FRED, undated-a) and where AvB is not offered. We also determined what yearly amount of AvB (to the nearest $250 if under $1,000 and to the nearest $1,000 otherwise) would maintain retention in cases where unemployment decreases (that is, for unemployment rates between 0 and 3 percent). This was determined by simulating the change in force size without AvB and with AvB in increasing increments of $250 until we reached $1,000 and then in increments of $1,000, and choosing the minimum AvB value where the change in force size was greater than or equal to 0. Interestingly, we found that opening the possibility of

TABLE 6.1
Effect of Different Values of AvB on Tracked Warrant Officer Aviator Retention Relative to Baseline of No AvB

Yearly AvB Amount ($)	Overall Change in Force Size (%)
5,000	2.6
10,000	3.1
15,000	3.8
20,000	4.7
25,000	5.7
30,000	6.2
35,000	7.1
40,000	7.9
45,000	8.8
50,000	9.3

NOTE: AvB amount is yearly and assumes a three-year contract. The yearly AvB amount for untracked aviators was $0. Simulated warrant officers are eligible for AvB beginning after the end of the ten-year initial ADSO until 13 years of AFS (including any enlisted service prior to entering the Army Aviation Branch) and from 19 to 22 years of AFS. See Table D.3 for results assuming a six-year initial ADSO.

TABLE 6.2
Effect of Changes in Unemployment Rate on Warrant Officer Retention With and Without AvB, Relative to a Baseline of No AvB and 4.1 Percent Unemployment

Unemployment Rate (%)	Change in Overall Force Size (%)	Yearly AvB Value That Maintains Retention (Unemployment Rate Decreases Only) ($)
0	−2.0	0
1	−1.5	0
2	−1.0	0
3	−0.5	0
4	0.0	NA
5	0.5	NA
6	1.0	NA
7	1.6	NA
9	2.6	NA
10	3.1	NA

NOTE: AvB amount is yearly and assumes a three-year contract. Simulated warrant officers are eligible for AvB beginning after the end of the ten-year initial ADSO until 13 years of AFS (including any enlisted service prior to entering the Army Aviation Branch) and from 19 to 22 years of AFS. See Table D.4 for results assuming a six-year initial ADSO. NA = not applicable.

a three-year contract with a bonus of $0 increases retention by enough to wipe out the retention changes from a decrease in the unemployment rate to 0 percent. Changes in retention are larger under a six-year ADSO, but the finding that opening the option of a contract without a bonus is enough to maintain retention stands (see Table D.4).

Table 6.3 reports the simulation results when civilian pay rises faster than military pay, with the change in force size reported relative to a baseline where the difference between civilian pay and military pay is set to its level in FY 2019 and where AvB is not offered. This situation is intuitively equivalent to a situation where the yearly military pay raise is smaller than the average pay raise in the civilian labor market. We simulated retention when no AvB is offered and determined the level of AvB for tracked aviators that would maintain retention relative to the baseline of no AvB, using the same method we used for unemployment. Similar to the unemployment case, we found that opening the option of a three-year contract with no bonus would raise retention by enough to wipe out the decrease in retention associated with up to a 4-percent rise in civilian pay. If civilian pay rises by 5 percent without AvB being offered, overall force size would fall by 2.5 percent relative to baseline. Offering $2,000 per year in AvB to tracked aviators only would maintain retention relative to the baseline. If civilian pay rises by 10 percent more than military pay, overall force size would fall by 4.7 percent relative to baseline if no AvB is offered and $20,000 per year in AvB would maintain retention. That is, even a relatively small amount of AvB for only tracked aviators (at only two-thirds the level offered in FY 2020) can nearly make up for a very large economic change when using no AvB as the baseline. Changes in retention are larger under a six-year ADSO, but the retention-maintaining values of AvB are lower; see Table D.5.

The previous two sets of simulations were calculated relative to a baseline where AvB was not offered. However, AvB has been offered on and off over the past five years (see Appendix B), and in FY 2020, a typical value for AvB was $30,000 over a three-year contract for tracked aviators (ALARACT 075/2019, 2019). This is because, based on discussions with our sponsor's office, AvB has been needed in recent years to ensure that the Army has a sufficient number of aviators to accomplish its operational goals rather than solely being

TABLE 6.3

Effect of Changes in Civilian Pay (Relative to Military Pay) on Warrant Officer Retention With and Without AvB, Relative to No AvB and Baseline Civilian Pay

Increase in Civilian Pay Relative to Military Pay (%)	Change in Overall Force Size When No AvB Offered (%)	Yearly AvB Value That Maintains Retention ($)
1	−0.5	0
2	−1.0	0
3	−1.5	0
4	−2.0	0
5	−2.5	2,000
6	−3.0	6,000
7	−3.5	10,000
8	−3.9	13,000
9	−4.3	17,000
10	−4.7	20,000

NOTE: AvB amount is yearly and assumes a three-year contract. Simulated warrant officers are eligible for AvB beginning after the end of the ten-year initial ADSO until 13 years of AFS (including any enlisted service prior to entering the Army Aviation Branch) and from 19 to 22 years of AFS. See Table D.5 for results assuming a six-year initial ADSO.

a tool to guard against changes in the external labor market. For practical purposes, if the Army is setting AvB levels to maintain retention relative to its current levels, then the baseline should include AvB at its most recent level. We therefore repeated the previous two analyses with a baseline where $30,000 yearly AvB over a three-year contract is offered to tracked aviators who have finished their initial ADSO and have either less than 14 or between 19 and 22 years of AFS. In these new analyses, we analyzed the effect of changes in the unemployment rate and changes in civilian pay when AvB is maintained at its baseline level and determined the level of AvB for tracked aviators that would maintain retention relative to the baseline of no AvB. The results are reported in Tables 6.4 (changes in the unemployment rate) and 6.5 (changes in civilian pay). We found that an increase of $14,000 per year (to $44,000 per year) in AvB for tracked aviators would maintain retention in the face of a decrease in the unemployment rate to 0 percent. We also found that an increase to $49,000 per year in AvB for tracked aviators would maintain retention in the face of a 5-percent increase in civilian pay and an increase of $36,000 per year (to $66,000 per year) in AvB for tracked aviators would maintain retention in the face of a 10-percent increase in civilian pay. Results under a six-year ADSO are reported in Tables D.6 and D.7; retention-maintaining AvB values are typically lower under the six-year ADSO than the ten-year ADSO.

Discussion of Findings on Retention Effects of AvB

In this chapter, we provided simulation results on the retention effects of varying levels of AvB and of economic changes with and without AvB. In Table 6.1, we provided information on how much overall force size increases under several different yearly amounts of AvB, which the Army can use as a tool for setting AvB amounts in the future. We then simulated how overall force size would change in the face of economic changes when AvB is or is not offered—both changes in the unemployment rate and changes in civilian pay. We found that opening the option of a three-year contract, even one not associated with a bonus, is associ-

TABLE 6.4

Effect of Changes in Unemployment Rate on Warrant Officer Retention With and Without AvB, Relative to a Baseline of $30,000 AvB for Tracked Aviators and 4.1 Percent Unemployment

Unemployment Rate (%)	Change in Overall Force Size When Tracked Offered $30k AvB (%)	Yearly AvB Value That Maintains Retention (Unemployment Rate Decreases Only) ($)
0	−2.2	44,000
1	−1.7	41,000
2	−1.2	37,000
3	−0.7	34,000
4	−0.1	31,000
5	0.5	NA
6	1.1	NA
7	1.6	NA
8	2.2	NA
9	2.8	NA
10	3.4	NA

NOTE: AvB amount is yearly and assumes a three-year contract. Simulated warrant officers are eligible for AvB beginning after the end of the ten-year initial ADSO until 13 years of AFS (including any enlisted service prior to entering the Army Aviation Branch) and from 19 to 22 years of AFS. See Table D.6 for results assuming a six-year initial ADSO.

TABLE 6.5

Effect of Changes in Civilian Pay (Relative to Military Pay) on Warrant Officer Retention With and Without AvB Increase, Relative to $30,000 AvB for Tracked Aviators

Increase in Civilian Pay Relative to Military Pay (%)	Change in Overall Force Size When Tracked Offered $30k AvB (%)	Yearly AvB Value That Maintains Retention ($)
1	−0.5	33,000
2	−1.0	37,000
3	−1.4	40,000
4	−2.0	44,000
5	−2.5	49,000
6	−3.0	52,000
7	−3.7	55,000
8	−4.1	59,000
9	−4.6	63,000
10	−5.1	66,000

NOTE: AvB amount is yearly and assumes a three-year contract. Simulated warrant officers are eligible for AvB beginning after the end of the ten-year initial ADSO until 13 years of AFS (including any enlisted service prior to entering the Army Aviation Branch) and from 19 to 22 years of AFS. See Table D.7 for results assuming a six-year initial ADSO.

ated with a retention increase large enough to wipe out a decrease in the unemployment rate to 0 percent or up to a 4 percent increase in civilian pay relative to military pay if the Army wishes to maintain retention relative to the level when AvB is not offered, and an increase of only $20,000 per year is required to maintain that level of retention in the face of a 10-percent rise in civilian pay. Larger bonus increases will be required if the Army wishes to maintain retention relative to a situation where warrant officers are already being offered $30,000 per year in AvB.

The fact that a relatively small increase in AvB can make up for a 10-percent civilian pay raise is surprising. Suppose that we were thinking about a warrant officer who entered the Army as an enlisted soldier at age 20, served five years before transitioning to a warrant officer, and had just finished their ten-year initial ADSO. As shown in Table C.3, the 70th percentile of civilian pay for 35-year-old veterans with one to three years of college is $68,782, so a ten-percent civilian pay raise is $6,878 per year every year until civilian pay peaks in their early 50s. This means that over the next 20 years in the civilian labor force, civilian pay would rise by over $137,000 relative to military pay, more than double the additional $60,000 of AvB over the three-year contract that we included in our simulations where the baseline is no AvB being offered. Even when we used a larger baseline of $30,000, AvB would only need to rise by $108,000 total to maintain retention in the face of a 10-percent civilian pay raise.

We believe three main factors can account for our results. First, AvB is targeted at warrant officers reaching two critical points in retention: at the end of their initial ADSO and at the point of retirement vesting. Warrant officers at both points are more likely to be sensitive to additional money: The end of the initial ADSO is the first point at which they are able to freely make the decision to leave the Army, and at the point of retirement vesting, they become eligible for a lifetime annuity. If AvB were offered at different YOS, the retention results might be substantially smaller. Second, our model included the value of the ability to make future choices and to reoptimize if circumstances change. An aviator who signs a three-year contract is, in our model, locked into the Army for three years and cannot reoptimize if circumstances change. In our model, there is some value to choosing to sign a three-year contract other than AvB (which is represented by the within-nest shock term), so even without the bonus, some warrant officers would choose to sign a three-year contract, raising overall retention. This is also part of the reason that larger increases in AvB are required to maintain retention when AvB is offered at baseline—the Army cannot take advantage of the value of opening the option of a three-year contract. Third, and related to our second point, the amount of emphasis that individuals in our model put on current versus future choices depends on their *discount factor*, which is a parameter that our model assumed rather than estimated. The discount factor, which falls between 0 and 1, is often thought of by economists as a measure of patience: Values close to 1 indicate that individuals care a lot about the future, and as the value shrinks, individuals care less about the future and more about the present. We assumed a relatively low discount factor of 0.91 in our analyses, which intuitively means that warrant officers in our model care substantially more about current choices than future choices and thus are probably weighing the bonus against a small number of years in the civilian labor force (rather than, for example, all the years they would spend in the civilian labor market). It is possible that alternative choices of the discount factor would change our results. That said, our discount factor assumptions were based on prior research where the discount factors were estimated for Army officers and enlisted personnel, as described earlier.

CHAPTER 7

Simulated Retention Effects of Increasing Service Obligation Length and of the Blended Retirement System

This chapter presents the simulation results for the retention effects of two previous policy changes: the increase in the Army Aviation Branch's ADSO from six to ten years and the BRS, because the potential retention effects of BRS have been a cause of concern for some Army Aviation force managers.

Increase in Service Obligation Length

As noted in Chapter 2 and Appendix A, aviators who entered aviation service in 2021 faced a ten-year ADSO following IERW, whereas previously, the initial ADSO was only six years. As part of our analysis, we simulated the retention response to the change in ADSO, holding all else equal, including tastes for active service among entering aviators. It is possible that a change in ADSO could affect the tastes for active service among entrants, so our results could overstate or understate the overall retention effect of a change in ADSO.

The results are shown in Figure 7.1, where the black line represents the Kaplan-Meier survival curve under the six-year ADSO, and the red line represents the predicted Kaplan-Meier survival curve under the ten-year ADSO. Retention increases substantially for both groups with the new ADSO: Commissioned officer retention is predicted to increase by 25.4 percent over the entire career, and warrant officer retention is predicted to increase by 14.9 percent over the entire career.

These results highlighted one important limitation of simulation results under a ten-year initial ADSO, which we highlighted in the previous chapter: We held constant the composition of entering aviators, even in the face of policy changes that might have changed which individuals would be likely to enter aviation service. Because of this limitation, Appendix D presents the results of all simulations under a six-year ADSO.

Impact of the BRS on Army Aviator Retention

The BRS went into effect on January 1, 2018. All personnel entering the military on or after that date are under BRS, and all personnel on hand and with fewer than 12 completed YOS on December 30, 2017, had the option of remaining under the legacy system or entering BRS.

BRS is so named because it is a blend of defined-benefit and defined-contribution pension systems, and it also includes a new pay, continuation pay. Table 7.1 compares key elements of BRS with the legacy system.

The defined-benefit part essentially continues the prior legacy retirement system but with two distinctly different features. It has a multiplier of 2 percent per YOS rather than 2.5 percent per YOS under the legacy retirement system. Also, at the time of retirement from the military with 20 or more years of AC service, a

FIGURE 7.1
Retention Effects of Increased ADSO Length

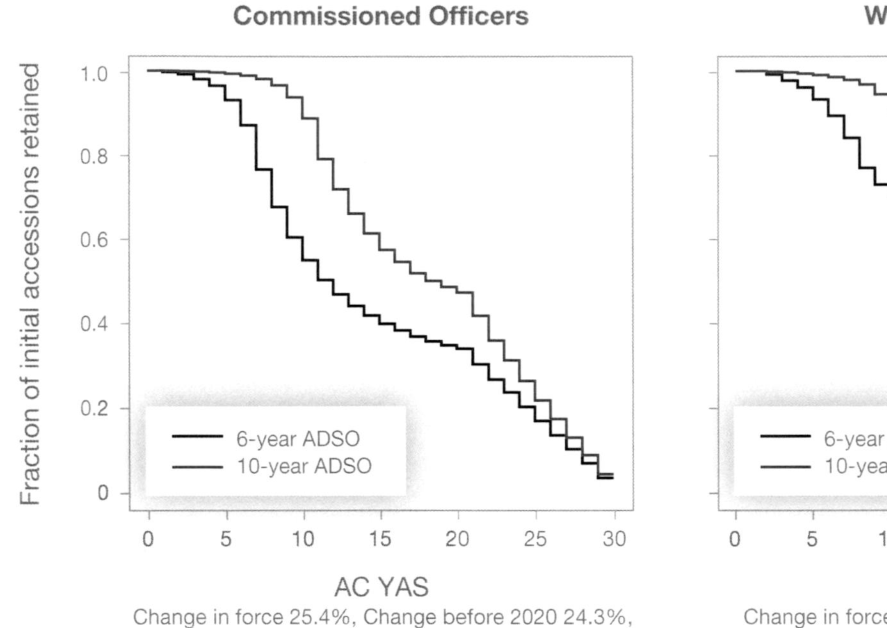

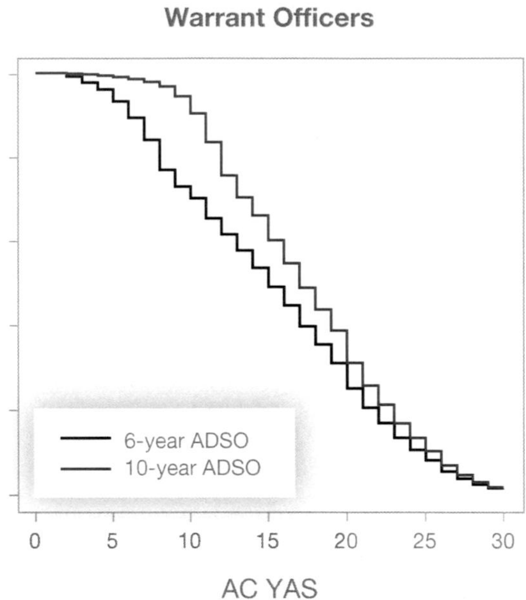

TABLE 7.1
Comparison of the Legacy and Blended Retirement Systems

Plan Element	Legacy	BRS
DB vesting	20 years	20 years
DB multiplier	2.5%	2.0%
DB payment working years		Full annuity or lump-sum option (50% or 25%); RC lump sum based on annuity from retirement age of 67
DC agency contribution rate		1% automatic, plus up to 4% matching (max = 5%)
DC contribution rate YOS		1%: entry plus 60 days until 26 YOS matching: start of 3 YOS through 26 YOS
DC member contribution rate		5% automatic; full match requires 5% contribution
DC vesting		Start of 3 YOS
Continuation pay multiplier (months of basic pay)		Minimum 2.5 for AC, 0.5 RC; with additional amount varying
CP YOS/additional obligation		At 12 YOS with 4 years of additional ADSO
Opt-in		Must be serving on January 1, 2018, and have fewer than 12 YOS (or a reservist with fewer than 4,320 points) as of December 31, 2017; opt-in period is January 2, 2018 through December 31, 2018

NOTE: Adapted from Asch, Mattock, and Hosek, 2017, Table 2.1. AC = active component; CP = continuation pay; DB = defined benefit; DC = defined contribution; RC = reserve component.

service member can elect to receive a lump sum based on the discounted amount of either 25 or 50 percent of the retirement benefits payable from retirement to the full social security age (67 for most people).

The defined-contribution part whereby DoD makes contributions to the plan is entirely new. It vests after the completion of 2 YOS, and DoD makes an automatic contribution of 1 percent and matches service member contributions up to an additional 4 percent when the service member contributes 5 percent. New service members are automatically enrolled to contribute 5 percent of their basic pay by default.[1] The defined-contribution funds are available at age 59.5. DoD matching continues through 26 YOS.

The final element of BRS, continuation pay, is a multiple of monthly basic pay. It is a one-time payment, payable between 8 and 12 YOS. It entails at least a three-year obligation and will be no less than 2.5 months of basic pay.

The following section shows how we simulated the impact of the BRS on retention under the default continuation pay for Army aviators and shows the steady-state levels of continuation pay required to sustain retention for warrant and commissioned officers. We explored how these retention-maintaining levels vary depending on the length of the ADSO and the availability of AvB.

How Will BRS Affect Retention?

The retention effects of BRS have been a cause of concern for some force managers because BRS decreases the defined benefit, which means that the draw of staying to 20 YOS is weaker than it previously was. Potentially offsetting this effect, service members vest in the defined-contribution system after 2 YOS and so have a portable retirement benefit under BRS that they own and control, even if they leave before 20 YOS. These aspects suggest that personnel are more likely to leave the military in mid-career years, and the military is more likely to lose their training, experience, expertise, and leadership. The risk of service members choosing to separate could be even greater if military pay lags behind civilian pay, external employment conditions improve, or both.

Despite these concerns, our analysis found that retention is unlikely to suffer substantially under BRS. Although the defined benefit is smaller, it is still 80 percent of what it was, which offers a strong draw to stay until 20 YOS to be eligible for the immediate payment of these benefits (that is, immediate receipt of the retirement annuity). Second, defined-contribution matching continues to 26 YOS, and the amount of the DoD match increases as basic pay increases. This is a significant contribution to the retirement nest egg, and the opportunity for future DoD matching contributions provides a draw for service members to stay. Third, the continuation pay multiplier can be adjusted depending on retention. Like bonuses and S&I pays, the continuation pay multiplier can be increased to ensure that retention goals are met and to draw service members through mid-career. Fourth, the opportunity to collect a lump-sum payment upon retirement can provide an additional draw for some service members.

In the case of Army aviators, we found that the minimum continuation pay multiplier generally does not sustain retention. However, the multiplier can be adjusted to fully restore retention. For example, the left panel of Figure 7.2 shows the retention of Army commissioned officer aviators when the continuation pay multiplier is 2.5. Under the minimum continuation pay, the steady-state decline in retention is 1.1 percent. If the continuation pay multiplier is raised to 9, then retention is fully restored to the baseline strength, albeit with slightly greater strength in mid-career and slightly less strength around the time of vesting in the defined-benefit element of the BRS at 20 YOS. This level of continuation pay required to restore officer retention is consistent with earlier findings, although smaller than the 10.85 optimized continuation pay for

[1] The default contribution rate was originally 3 percent but was later raised to 5 percent on October 1, 2020 (Thrift Savings Plan [TSP] Bulletin 20-7, 2020).

FIGURE 7.2
Army Commissioned Officer Aviator Retention Under BRS

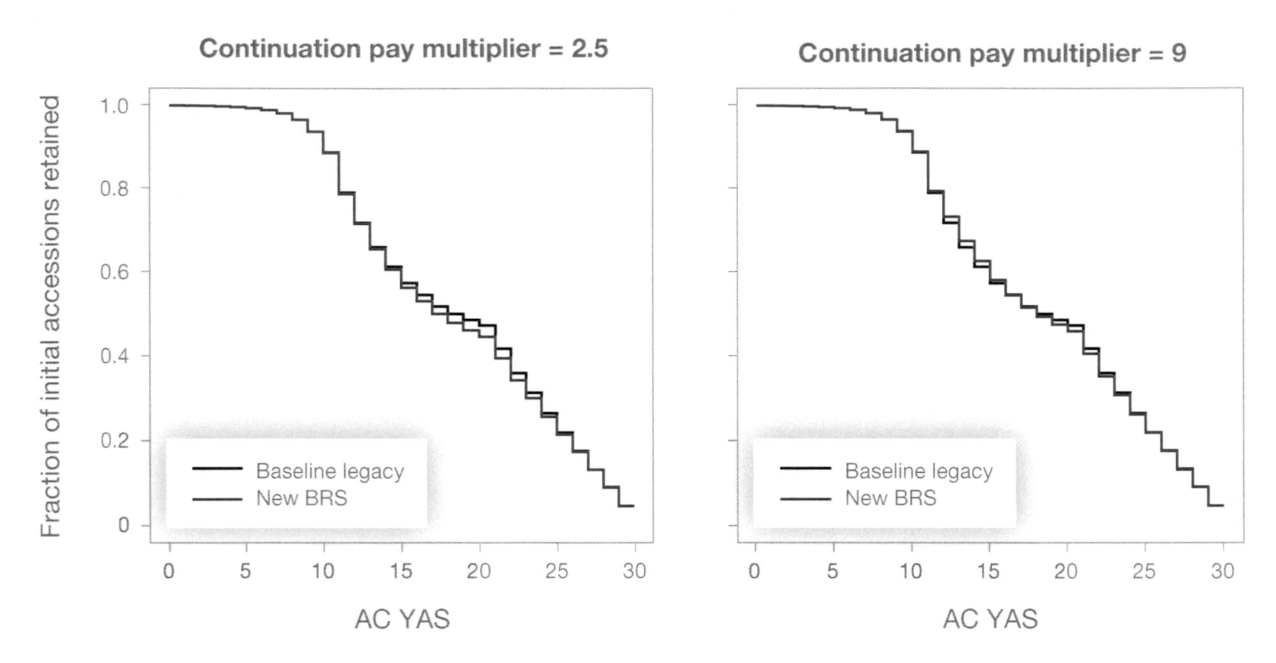

all Army commissioned officers reported in Asch, Mattock, and Hosek (2017). This may be because the taste distribution is higher for commissioned officers who are aviators, and thus, a lower level of continuation pay is required to restore retention.

Similarly, warrant officer retention declines under the minimum continuation pay, although the steady-state decline will vary depending on whether or not the AvB is offered. If AvB is not offered, as is depicted in Figure 7.3 (versus Figure 7.4, which depicts the same population with AvB), then the decline under the minimum continuation pay is 0.8 percent, and retention can be restored by raising the continuation pay multiplier to 4.5. Under a scenario where AvB is consistently available, the steady-state decline in retention when the minimum continuation pay is offered is only 0.2 percent, and retention can be brought up to the baseline level when the continuation pay multiplier is raised to 3.25. These continuation pays fall within the range of optimized continuation pays for Army service members reported in Asch, Mattock, and Hosek (2017), ranging from 2.39 for enlisted service members and 10.85 for commissioned officers. Continuation pays for warrant officers were not calculated.

Summary

In this chapter, we simulated the impact of two previous policy changes on the retention of Army Aviation warrant officers and commissioned officers. First, we simulated the effect of changing the length of the ADSO incurred by officers selected for initial entry flight training from six years to ten years. We found that retention would significantly increase for both warrant officers and commissioned officers. Second, we simulated the effect of changing from the legacy retirement system to the BRS. We found that in the steady state, when all officers are under the BRS for their entire career, the minimum continuation pay multiplier of 2.5 is not sufficient to sustain retention for either warrant officers or commissioned officers. However, if the continuation pay multiplier is raised to 9 for commissioned officers, retention will be fully restored. The required continuation pay multiplier for warrant officers depends on whether or not AvB is available. If it is,

FIGURE 7.3
Army Warrant Officer Aviator Retention Under BRS Without AvB

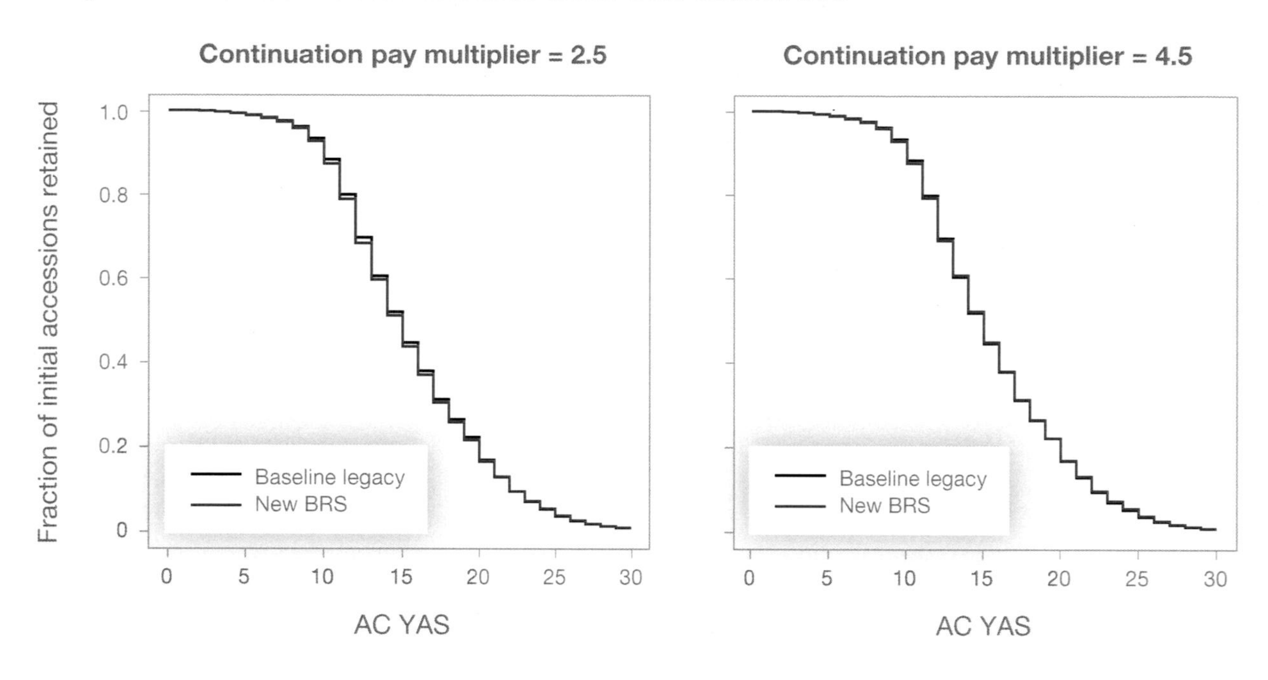

FIGURE 7.4
Army Warrant Officer Aviator Retention Under BRS with AvB

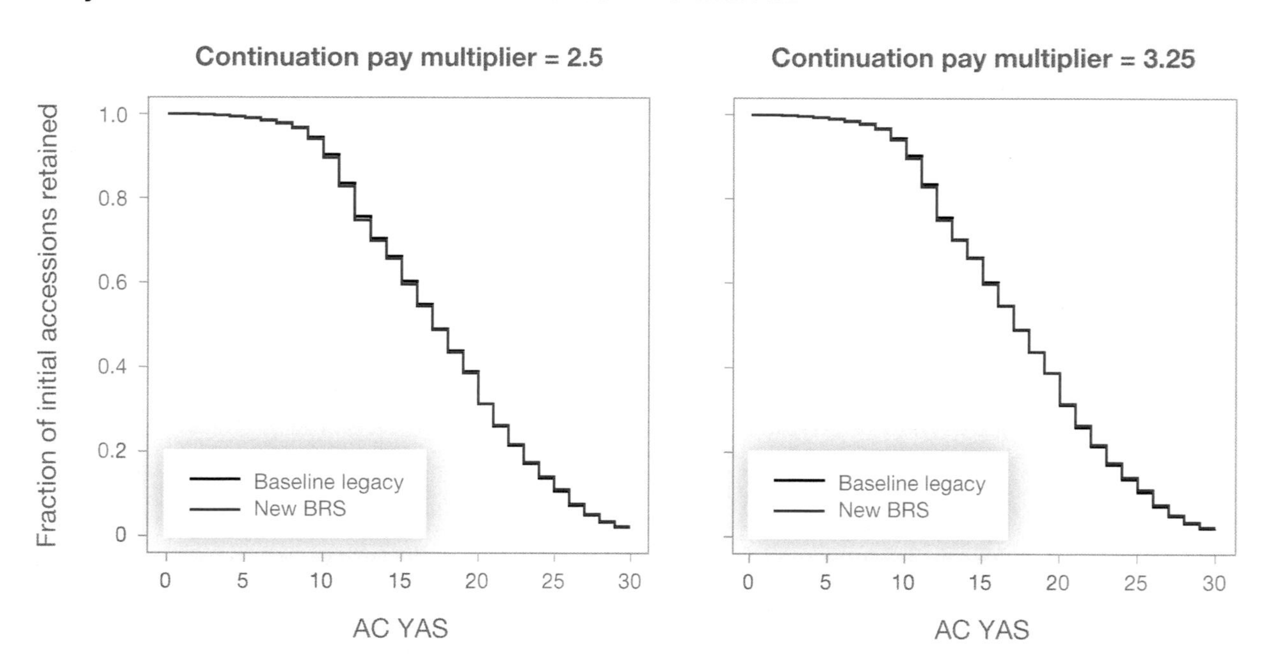

then retention will be restored with a multiplier of 3.25. If it is not, then the required multiplier will be 1.25 units higher at 4.5.

CHAPTER 8

Findings and Conclusions

To modernize its S&I pays for aviators, the Army is considering proposals that would make AvIP contingent on achieving specific career milestones for aviators, such as achieving pilot-in-command status or on enrolling in PME. Making pays contingent on these career milestones could provide increased retention incentives for those who achieve those milestones. However, for any reform proposal to move forward with the other services, OSD, and Congress, analysis is needed on how it would affect AC aviator retention of both warrant officers and commissioned officers, as well as its cost and performance. Beyond S&I pays based on the achievement of milestones, the Army is also concerned about other factors that could adversely affect aviator retention and about how changes in aviator S&I pays might address drops in retention. These factors include strong civilian demand for pilots, reductions in the real value of AvIP because of the eroding effects of inflation, and the move to the BRS. We used Army personnel data to estimate DRMs for Army warrant officer and commissioned officer aviators and found that predicted retention with our models fit the observed data well. We then used the model estimates to simulate the effects of alternative policies on aviator retention, including setting AvIP based on the achievement of career milestones. We summarize the key findings and conclusions in this chapter.

Limitations

Before summarizing our key findings, we note that our work has several limitations. First, our simulations of the retention effects of AvIP based on career milestones depended on our assumptions regarding the timing and probability of achieving career milestones. While we based these assumptions on observed values in the Army data, we also performed sensitivity checks where we varied the assumptions regarding timing and probability and found that our estimates of which values of AvIP for each milestone would maintain retention did vary a small amount depending on which assumptions we used. Second, we did not observe or simulate the incentive to acquire additional human capital and the effort required to achieve aviation milestones because it would require too much computing power and too many assumptions (see Asch et al. [2020] for a description of the required assumptions and computing power). Third, our assessment of the retention effects of different policies, such as increasing the initial ADSO, did not consider how these policies would affect the taste distribution of the entry cohort and thereby affect subsequent retention. For example, raising the ADSO could result in fewer entrants with a higher taste for active service with a higher propensity to stay in service. Our retention estimates did not consider these second-order effects. However, we performed sensitivity analyses for our simulations, where we reran all simulations assuming a six-year initial ADSO. Those results are included in Appendix D. In general, retention responses to changes in pay are typically larger under the six-year initial ADSO because more aviators are free to stay or leave under this policy than under the current policy.

Setting AvIP Based on Career Milestones Will Require Higher AvIP Levels

The Army provided us with a set of prototype milestone-based AvIP values as a starting point. These prototype milestone-based values were similar to the YAS-based AvIP values that the Army currently uses. For example, under current policy, monthly AvIP for aviators with 2 YAS equals $200. Under the prototype, personnel who achieve pilot status, typically at around 1.5 YAS (rounded up to 2 YAS in our model), would receive $200 monthly.

The advantage of setting AvIP based on milestone achievement is that it increases the incentive of aviators to achieve the milestones. It also increases retention incentives more for those who achieve the milestone than for those who do not achieve the milestone. A disadvantage of this approach is that not all aviators who remain in the Army achieve the key career milestones. For example, only 65 percent of warrant officer aviators who stay until 10 YAS become senior aviators and would receive the $1,000 monthly AvIP. In contrast, under current policy, all aviators at 10 YAS would receive the $1,000 monthly AvIP. This approach not only introduces inequity between aviators at a given YAS, but also it means that we would expect the retention effect of AvIP to be smaller because the likelihood of being paid a higher AvIP is lower under the alternative policy.

Our analysis confirmed this intuition regarding the retention effects of basing AvIP on milestones. We found that switching to setting AvIP values based on career milestones that equal the current policy, as under the Army's prototype, would reduce retention (as measured as the change in the size of the aviator inventory) by approximately 1.0 percent for commissioned officers and 1.9 percent for warrant officers. We also estimated that it would reduce AvIP cost to the Army, both because some aviators under the new policy at a given YAS would receive lower AvIP and because retention would be lower (so AvIP would be paid to fewer personnel).

We then searched for potential values of AvIP which would maintain retention but still be based on milestones—a retention-maintaining AvIP schedule. We found that the Army would need to offer substantially higher values of AvIP than the prototype if they wish to maintain retention, especially to senior and master aviators (though we note that the values we identified as sustaining retention may not be unique). Maintaining retention with AvIP based solely on milestones would also increase AvIP cost relative to current policy. Costs increase under the alternative policy because higher dollar values of AvIP are required to sustain retention, and this cost increase more than offsets the lower costs that occur because not all aviators at a given YAS would receive higher AvIP, as under current policy.

The retention-maintaining AvIP values would not be the same for commissioned officers and warrant officers. Under current policy, AvIP is the same dollar amount for both groups. The values that sustain retention when AvIP depends on milestones differ. In particular, the values that sustain retention for commissioned officers would be higher than those for warrant officers, owing to different retention behaviors of each group and differences in their responsiveness to changes in compensation. Different AvIP values between warrant officers and commissioned officers would introduce a source of inequity that currently does not exist. However, should the Army decide to reintroduce equity and set the same AvIP levels for both groups, based on milestones, our analysis indicated that retention would not be sustained for one group. For example, should the Army set warrant officer AvIP to the commissioned officer level, warrant officer retention would exceed the level achieved under current policy, costs would increase, and warrant officers would receive economic rents, meaning they would receive higher pay than required to sustain their retention relative to the baseline.

We did not analyze whether basing AvIP on the achievement of career milestones would increase the incentive of aviators to acquire more human capital and achieve those milestones because such an analysis was beyond the scope of the project. However, we did investigate the effect on retention of increasing the likelihood of achieving milestones among all aviators, which would likely be the result of increased human capital accumulation. We found that the values of AvIP based on milestones would not need to increase as much to maintain retention as they would if the probability of milestones remained constant. The increase in cost would also be lower if the probability of achieving milestones increased. Thus, to the extent that basing AvIP on milestones increases the likelihood that a pilot acquires the human capital and achieves the milestone, the increase in cost and the degree to which AvIP must be higher will be reduced.

Setting AvIP Based on Career Milestones Can Increase the Incentives for Higher-Ability Pilots to Remain in Service

Another potential advantage of basing AvIP on milestone achievement is the potential effect on the retention of higher-ability personnel. To the extent that higher-ability aviators achieve these milestones faster, basing AvIP on milestones also increases the incentive for higher-ability aviators to stay in the Army. We found that neither the Army's prototype AvIP nor retention-maintaining values of AvIP substantially raise the average ability of the aviator force, nor do they substantially raise ability at higher ranks. The Army's prototype AvIP and the retention-maintaining AvIP apparently offer insufficient rewards to substantially raise retention among its highest-ability personnel.

However, we identified two modifications to the AvIP schedule that do increase retention. In both modifications, personnel who achieve the milestones sooner receive higher AvIP payments than personnel who achieve milestones later. Under the first variant, the value of AvIP values increases in a way that is proportional to the retention-maintaining AvIP schedule. Under the second variant, AvIP values double at each successive milestone. Under these proposals that increase the reward for achieving milestones, especially for higher-ability aviators, we found that the retention of higher-ability personnel increases. We also found that ability sorting improves—that is, under these proposals, more high-ability personnel are retained at later milestones than under the current baseline system.

Retention Increases When the Initial Service Obligation Is Increased

Our models were estimated for aviators who entered the Army under a six-year initial ADSO policy, but our main results were produced using a ten-year initial ADSO because the Army moved to a longer ADSO at the beginning of FY 2021. As expected, we found that increasing the initial ADSO to ten years substantially increases retention over the entire aviator career, not just during the first ten years. For commissioned officers, force size in the steady state is predicted to rise by 25.4 percent, while the increase for warrant officers is predicted to be 14.9 percent. These increases are likely because of service members being four years closer to vesting in the defined-benefit retirement annuity, while the difference between commissioned officers and warrant officers is due, in part, to differences in how much emphasis individuals put on current versus future income (i.e., discounting or time preference)—commissioned officers value the retirement annuity more. As noted above, these estimates did not consider how such a policy would affect the taste mix of entrants. However, we did conduct sensitivity analyses of our results, presented in Appendix D, to show how our results regarding the effects of compensation changes on aviator retention would change under an initial ADSO of six years rather than the ten-year assumption.

Increases in AvB Can Offset the Effects of Increased Civilian Demand

We also examined retention responses among warrant officers to AvB, a large retention incentive that is typically tied to a three-year ADSO. We provided a table of retention responses to various amounts of AvB, which the Army can use as a reference when setting AvB to target various retention targets. We first examined a base case where no AvB is offered. We found that offering $20,000 per year in AvB to tracked warrant officers only can offset Army Aviation retention in the event of implausibly large shocks to external opportunities (such as a 10-percent increase in civilian pay relative to military pay) and that merely opening the option of a three-year contract with no bonus will increase retention enough to offset large changes in the unemployment rate (such as a drop to 0 percent unemployment). That is, a small level of AvB could restore retention to the levels observed at unchanged levels of civilian demand. Small levels of AvB are able to offset the less extreme of our shocks when a $30,000 AvB per year is offered in the base case; an additional $14,000 ($42,000 total) can offset the effect of a drop in the unemployment rate to 0 percent; and an additional $19,000 ($57,000) can offset a 5-percent increase in civilian pay. However, a larger AvB increase (of $36,000 per year or $108,000 total) is needed to offset a 10-percent increase in civilian pay.

Increased Continuation Pay Can Offset Adverse Retention Effects of the BRS

A key element of the BRS is continuation pay, a new pay offered to service members with between 8 and 12 YOS that is equal to a congressionally mandated minimum of 2.5 times monthly basic pay. Continuation pay, together with the defined-contribution plan, known as the TSP, are intended to offset any adverse effect on retention of reducing the value of the legacy defined-benefit system under BRS. Earlier research (Asch, Hosek, and Mattock, 2017) found that the default 2.5 multiplier is sufficient to sustain Army enlisted retention but too low to sustain officer retention.

Our simulations indicated that under BRS, the default continuation pay multiplier of 2.5 is not sufficient to sustain the level of retention of either commissioned or warrant officer aviators observed under the legacy retirement system. Those simulations also indicated that aviator retention can be sustained by raising the multiplier to 9 for commissioned officers, raising to 3.25 for warrant officers if AvB is offered, or raising to 4.5 for warrant officers if AvB is not offered.

Conclusions

Our analysis indicated that setting AvIP based on career milestones could sustain Army aviator retention. But doing so would increase AvIP costs and overall personnel costs to the Army because AvIP amounts need to be higher than current levels (if the Army wants to replicate current retention levels) to offset the fact that not all aviators achieve the key career milestones and so would not be eligible for the higher AvIP amounts associated with reaching those milestones. We found that an increase in the likelihood of achieving milestones would mean that costs would increase by less. Greater retention of higher-ability aviators and improved ability sorting could potentially offset the higher costs, although these improvements only occur when AvIP values sufficiently reward the achievement of milestones by higher-ability personnel.

We also found that while the Army increased AvIP values in 2020, after more than 20 years of decreasing real value, this increase was insufficient to sustain retention. Our model predicted a drop in retention where the real value of AvIP was maintained since 1998. We also found that the Army's policy to set continuation pay at the minimum value under the BRS results in a decline in aviator retention. The Army could sustain retention under BRS for aviators by selecting higher continuation values than the minimum for both commissioned officers and warrant officers. Finally, we found that improvements in the civilian economy, both in terms of the unemployment rate and civilian pay relative to military pay, adversely affect warrant officer aviator retention, but increases in AvB can offset these effects.

To summarize, making S&I pays contingent on achievement of career milestones could provide incentives for the development of valuable human capital for the Army. This reform could be designed to not only sustain retention but also to target compensation to individual qualifications and talent. However, for any reform proposal to move forward with the other services, OSD, and Congress, analysis is needed on how the reform would affect AC aviator retention of both warrant officers and commissioned officers, cost, and performance measured in terms of milestone attainment and the retention of higher-ability aviators. The analysis presented in this report is a step toward addressing that need.

APPENDIX A

Aviation Careers

This appendix describes aviation careers. We first present a detailed description of aviation careers. We then present detailed distributions of the timing of achievement of the aviation career milestones described in Chapter 2.

Army Aviation Career Milestones and Related Performance

Overview of the Army's Aviation Unit Structure

Army Aviation units consist of the Regular Army (RA), ARNG, and USAR. The Army estimates 3,750 of 5,000 aircraft and personnel are assigned to Modified Table of Organization and Equipment (MTOE) units, which are tactical units that can be deployed around the world to conduct Army operations ranging from combat to peacekeeping. The remaining fixed- and rotary-wing aircraft and personnel are assigned to Table of Distribution and Allowances (TDA) units, which are organizations that do not deploy, such as the Army Aviation Branch's training center at Fort Rucker, Alabama, or certain elements of the ARNG in each state (Holistic Aviation Assessment Task Force, 2016). MTOE formations are organized in one of seven brigade-level types of units, the most common being the combat aviation brigade (CAB) and the enhanced combat aviation brigade. Some unit designs include the theater aviation brigade, special operations aviation regiment, and theater airfield operations groups designed to provide airfield management in austere or combat environments (Holistic Aviation Assessment Task Force, 2016).

Table A.1 describes the occupations of officers in the Army Aviation Branch. RA CABs are generally located and organized in one location with subordinate elements, unlike USAR CABs that are organized into two enhanced CABs with subordinate units down to the company level dispersed around the country. This is done mainly for recruiting and retention purposes; however, USAR units are commanded by their organic higher headquarters (AR 600-105, 2020). For example, a USAR company located in Texas is commanded by

TABLE A.1
Aviation Officer Area of Concentration MOS Identifiers

Aviator Position	MOS Code	MOS Title
Officer-grade aviator	15A	Aviation General
	15B	Aviation Combined Arms Operation
	15C	Aviation All-Source Intelligence
	15D	Aviation Maintenance Officer
	67J	Aeromedical Evacuation
Warrant officer aviator	150 Series	Non-Aviator, Technical Expert
	151 Series	Non-Aviator, Maintenance
	152–155 Series	Aviator Specialties

SOURCE: Features data from Department of the Army Pamphlet 600-3, 2022.

its authorized battalion-level headquarters, even though that headquarters is based in Colorado. RA operates about 53 percent of the Army's MTOE rotary-wing aircraft, and the USAR and ARNG operate the rest (Holistic Aviation Assessment Task Force, 2016). Similar to the USAR, ARNG aviation is geographically dispersed around the country; however, ARNG units are subdivided below company level down to the detachment level, expanding further the separation of forces. It is important to note that unlike the USAR, ARNG subordinate units do not receive day-to-day missions from their higher headquarters; instead, daily missions are assigned by each state's leaders.[1] Personnel authorized to fly in Army aircraft are assigned to these units and are part of the rated inventory or are rated acquisition corps officers with a pilot status code of 1 in a valid position. This includes all commissioned officer aviators in the grade of warrant officer 1 (WO1) through colonel who have completed qualification, training, evaluation, and currency requirements adherent to Army regulations for the aircraft to be flown or are performing duties (AR 600-105, 2020).

Linkage Between Unit Assignment and Career Progression

In general, an aviation company is equipped with either manned rotary-wing or fixed-wing platforms or unmanned aircraft systems, a basic structure for the CABs. The set of Army rotary-wing platforms is composed of the Utility Helicopter (UH)-72 LAKOTA, UH-60 Black Hawk, Cargo Helicopter (CH)-47 Chinook, and Attack Helicopter (AH)-64 Apache (Department of the Army Pamphlet 600-3, 2022). A chart of company mission versus aircraft type is provided in Figure A.1.

Company-level organizations are combined at the battalion level and organized into CABs and theater aviation brigades. These units are manned with officers, warrant officers, non-commissioned officers, and enlisted soldiers (AR 600-105, 2020).

Based on discussions with our sponsor's office, assigned commissioned officers and warrant officers are categorized into various areas of concentration—most are aviators. This unit structure links directly to career development and progression models developed by U.S. Army Training and Doctrine Command.

FIGURE A.1
Army Aviation Company Mission and Aircraft Type

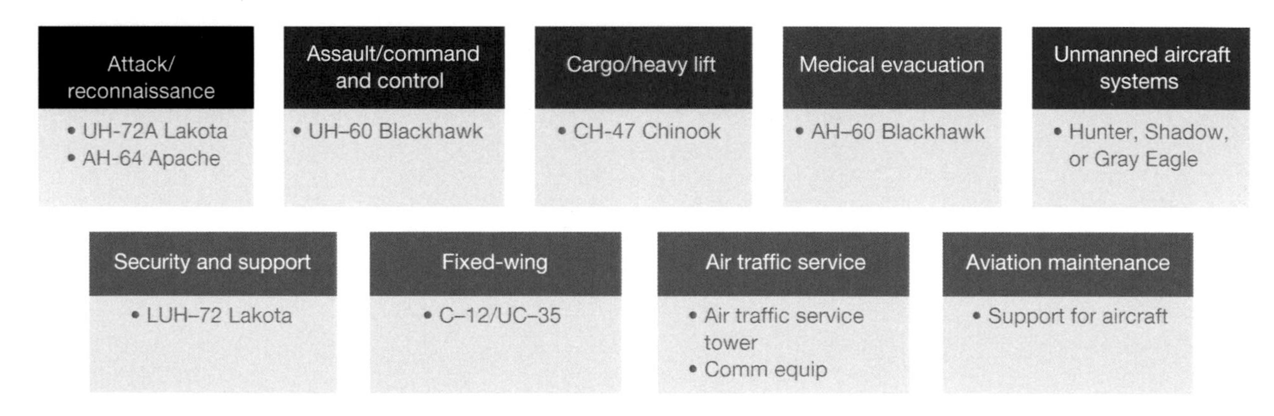

SOURCE: Features information from Department of the Army Pamphlet 600-3, 2022.
NOTE: Comms equip = communications equipment.

[1] This situation complicates the ability of personnel to gain access to critical training, ultimately leading to challenges reaching required benchmarks and milestones across the training domain. Training in the ARNG is executed with those commanders having no authority to train or direct units outside their state until the units have been mobilized and are under federal authority.

Aviator career development is based on operational experience and training, institutional education, and self-study and development.

Career progression depends on time in grade, duty performance, experience, schooling, and skills acquired, with minimum time-in-grade requirements for progression to the next rank set by the Army. Aviators are considered for promotion when they have achieved the minimum time in grade. Duty performance is the single most important factor in selection to progress to the next-higher rank. Operational experience is also a key factor in progression, particularly in the early- and mid-career development periods. Operational assignments that develop tactical and technical expertise are important at these stages. Broadening assignments help develop competencies beyond tactical and technical aviation expertise, becoming increasingly important during the mid-career and later development periods. Schooling and the performance of critical or unique skills can also enhance potential for progression. Certain PME courses are required for advancement to the next rank (Department of the Army Pamphlet 600-3, 2022).

In general, the first goal for the aviator is to establish a solid foundation of aviation tactical and technical expertise, then expand their operational- and strategic-level competence with respect to the Army and joint, interagency, intergovernmental, and multinational environments.

Commissioned officer aviators and warrant officers must achieve designation as an Army aviator at the beginning of their career and must undergo annual flight proficiency and aeromedical assessments to maintain the rating. Aviators undergo readiness-level training and certification upon arrival at each new flight unit to confirm their competency levels to perform as a crew member in their assigned aircraft. Aviators must achieve pilot-in-command status to be entrusted as the aircraft commander in an aircraft crew. To continue to receive aircrew incentive pay, commissioned officer aviators and warrant officers must accrue mandated benchmarks of operational flying duty credit through assignment to positions designated as operational flying positions (Army Regulation 600-105, 2020). Aviators must understand all aspects of integrating aviation platforms, systems, and units into effective air-ground operations. Commissioned officer aviators including warrant officers must understand the fundamentals of airspace management and air traffic service requirements as well. This includes compliance with airspace regulations of the Joint Force Airspace Control Authority, political host nation requirements, International Civil Aeronautics Organization guidelines, FAA regulations, and other pertinent regulations of airspace governing agencies as applicable, in addition to maintenance standards, processes, and procedures (AR 95-1, 2018).

Aviator career development (Figure A.2) and progression is measured and based on operational experience and training, institutional education, and personal development. Aviators must undergo readiness-level training and certification upon arrival at their initial and subsequent duty stations, changing on average every three to four years (Department of the Army Pamphlet 600-3, 2022).

They are required to confirm their competency to perform as a crew member in their assigned aircraft through training, their record of performance, and evaluations. To maintain incentive pay, commissioned officer aviators and warrant officers must accrue mandated thresholds of total operational flying duty credit through assignment to positions designated as operational flying positions (Department of the Army Pamphlet 600-3, 2022).

Commissioned Officer Aviator Duty Description

This section describes the career paths of commissioned officer aviators. Material in this section is based on discussions with our sponsor's office.

The aviation commissioned officer is first an expert aviator but also is responsible for the coordination of aviation operations from maintenance to control tower operations to tactical field missions. Each commissioned officer must undergo professional development over the course of their career, including institutional training, successfully serving in assignments in key development billets, serving as part of the joint staff or

FIGURE A.2
Aviator Career Development

Early career development

focuses on developing tactical expertise in employment and sustainment as part of the air-ground, integrated, and combined arms team and technical expertise in the operation of specific systems

Mid-career development

continues to refine tactical and technical expertise and focuses on developing competency in operational-level employment and sustainment of aviation and integration into joint, interagency, intergovernmental, and multinational environments

Advanced career development

focuses on strategic-level employment and sustainment of aviation and integration into the force

SOURCE: Features information from Department of the Army Pamphlet 600-3, 2022.

part of cross-service joint commands, completing broadening tours of duty with government agencies or commercial industry, and completing academic self-development.[2] Figure A.3 provides a description of commissioned officers' career paths.

Lieutenants must complete training as a commissioned officer aviator starting with IERW,[3] which is initial preflight training through primary and instrument qualification phases. During training they are taught basic warfighting skills and are certified on one of three assigned aircraft. Lieutenants are generally assigned to jobs that allow the officer adequate opportunity to develop flight experience and troop-leading skills over the first two years and to attain pilot-in-command status. Responsibilities of an aviation lieutenant may include

- attending Basic Officer Leaders Course and IERW training
- coordinating the employment of soldiers and aircraft from platoon to battalion and higher, in U.S. and multinational operations, and providing aviation coordination.

Around the fourth year of active service, lieutenants are promoted to the rank of captain. The primary focus upon promotion to captain is successfully completing command at the company level for two years. During this time, they complete related Army Aviation Branch career course training (i.e., the Captain's Career Course). After this training, between their fifth and eighth YOS, they transition from a 15A area of concentration to 15B. Responsibilities of an aviation captain may include

- commanding at the company level
- as a staff officer, coordinating employment of aviation soldiers at all levels of command, from company to division level and beyond, in U.S. and multinational operations
- developing plans and executing unique aviation missions
- instructing aviation skills at service schools and combat training centers
- serving as an aviation advisor to other units, including ARNG and USAR organizations.

Promotion to major typically occurs around the tenth year, following successful command and related branch career course attendance. Majors are asked to fulfill the requirement to complete intermediate level

[2] Positions listed help provide a continuous cycle of education, training, selection, experience, assessment, feedback, reinforcement, and evaluation, which helps to encourage officer development throughout career progression.

[3] Includes 2nd (Second) and 1st (First) Lieutenant ranks.

FIGURE A.3
Commissioned Officer Aviator Career Path

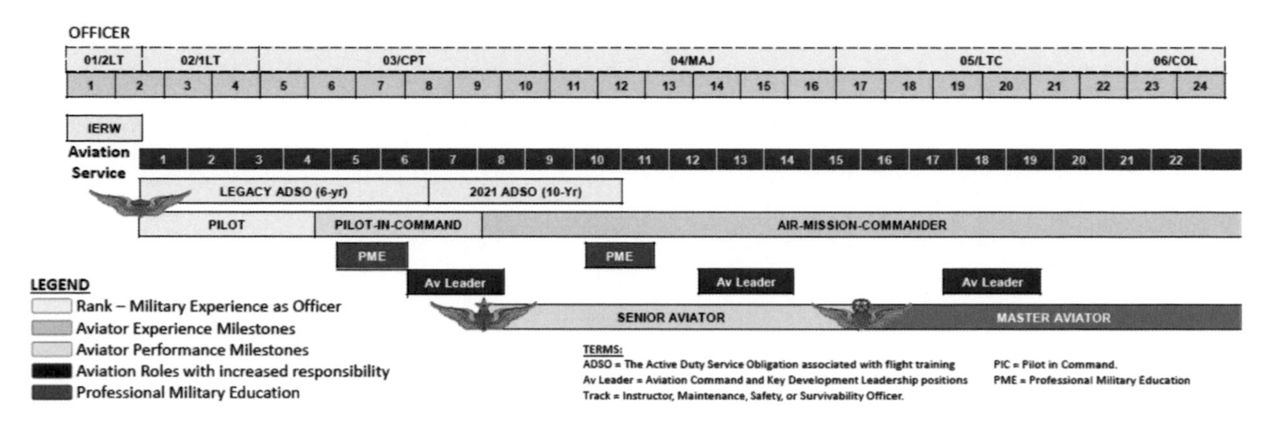

SOURCE: Created for the authors by the U.S. Army Aviation Center of Excellence, April 13, 2022.

education, a type of PME, at the Command and General Staff College, a sister service–equivalent institution (Navy, Marine Corps, or Air Force), or schools in other nations for consideration to be promoted. Regardless of their area of concentration, majors seek to serve in key development assignments, as a battalion- or brigade-level executive officer, as a division-level operations officer, or as a commander (position for majors). Responsibilities of an aviation major may include

- attending Command and General Staff College, a sister service–equivalent (Navy, Marine Corps, or Air Force), or schools in other nations for consideration to be promoted
- serving in key development assignments
  - battalion- or brigade-level executive officer
  - division-level operations officer
  - commander
  - two to three years in a Joint or Army Staff position
  - combat training center, Reserve Officer Training Corps, or U.S. Army Recruiting Command
  - two-year Training with Industry

Promotion to lieutenant colonel typically occurs around 16 YOS. As a lieutenant colonel, serving as a battalion-level commander in an aviation-coded position for two years is a highly competitive selective process and critical to being promoted to colonel. Lieutenant colonels are strongly encouraged to complete advanced PME, although no specific military education requirements exist. Responsibilities of an aviation lieutenant colonel may include

- commanding at the battalion level
- completing PME such as that offered through
  - U.S. Army War College Distance Education Course
  - Army's Pre-Command Resident Course and Aviation Pre-Command Course
- serving in assignments such as
  - two to three years with U.S. Army Training and Doctrine Command, U.S. Army Recruiting Command, or U.S. Army Forces Command
  - CAB deputy commander, executive, or operations officer.

A colonel's primary focus during the first three to four years is broadening their knowledge of sustainment of warfighting, training, and staff skills, along with improved leadership, managerial, and executive skills. The most competitive key developmental assignment sought is brigade commander. Colonels are also required to acquire key development time while serving as senior-level joint duty, division, corps-level officers for the Army. Responsibilities of an aviation colonel may include

- commanding at the brigade level
- serving in assignments such as
  - primary on the division or corps staff
  - U.S. Army Aviation Center for Excellence director
  - two-to-three-year Combat Training Center senior aviation observer assignment
  - two-to-three-year Joint or Army Staff assignment
  - U.S. Army Training and Doctrine Command or Army Futures Command capabilities manager.

Warrant Officer Aviator Duty Description

This section describes the duties and career progression of warrant officer aviators. Material in this section is based on discussions with our sponsor's office.

Warrant officer aviators make up over 60 percent of pilots and aviation support officers across Army formations operating and sustaining aircraft in tactical and non-tactical conditions. Warrant officers fill a unique role as the technical and tactical experts of the Army Aviation Branch, providing long-term continuity of service within both conventional and special operations aviation units. Warrant officers operate aircraft in all meteorological conditions, both day and night, and are responsible for coordinating, conducting, and directing all types of single service and joint combat, combat support, and sustainment operations. Figure A.4 provides a description of the warrant officer career path.

At the beginning of their aviation career, while at the grade of warrant officer flight training and WO1, warrant officers must attend the Warrant Officer Candidate School, IERW and Warrant Officer Aviator Basic Course, and the Survival, Evasion, Resistance, and Escape Course. WO1s receive appointment upon successfully completing MOS certification courses and graduation from Warrant Officer Aviator Basic Course.

FIGURE A.4
Warrant Officer Aviator Career Path

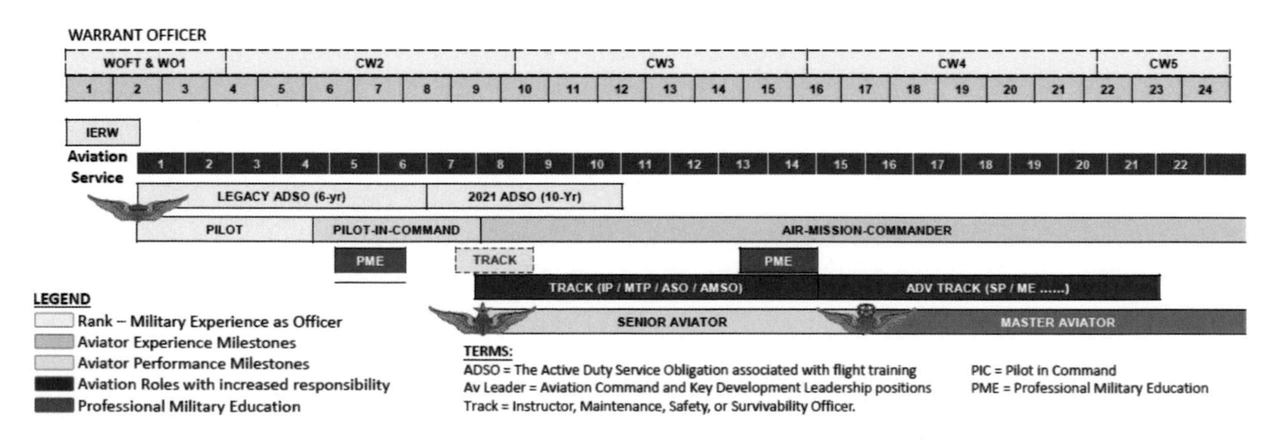

SOURCE: Created for the authors by the U.S. Army Aviation Center of Excellence, April 13, 2022.

Around the two-year mark, WO1s are promoted to chief warrant officer 2 (CW2) and serve as intermediate-level technical and tactical experts who perform the primary duties that support operations from crew to team to battalion levels. Responsibilities of a CW2 may include

- attaining pilot-in-command status
- completing career tracked training courses for safety, maintenance test pilot (MTP), mission survivability officer, and instructor pilot
- potentially, volunteering for assessment into Army Special Operations Aviation training
- serving in company-level assignments as pilot, aviation life support and survivability equipment officer, safety officer, instructor pilot, or MTP (e.g., assignments might include U.S. Army Training and Doctrine Command Instructor)
- attaining the Army's senior aviator badge.

A chief warrant officer 3 (CW3) is required to perform duties as an advanced-level technical and tactical expert, primarily supporting levels of operations from up to battalion and brigade levels. Responsibilities of a CW3 may include

- attending Warrant officer aviator Advanced Course no later than one year after promotion
- completing advanced tracked training as an MTP, aviation safety officer, or senior instructor pilot or instrument flight examiner
- attaining air mission commander
- serving in assignments such as
  - warrant officer recruiter
  - flight lead
  - Army Special Operations Aviation Command.

A chief warrant officer 4 (CW4) is a senior-level technical and tactical expert who serves at the field-grade level as a senior aviator and senior staff officer with a primarily support role at the battalion, brigade, division, corps, and echelons above corps levels. Responsibilities of a CW4 may include

- attending the Warrant Officer Staff Course not later than one year after promotion to CW4 and completing the course prior to promotion to chief warrant officer 5 (CW5)
- performing duties as squadron- and battalion-level aviation safety officer, standardization instructor pilot, maintenance test flight examiner (MTFE), aviation mission survivability officer, master gunner, or in Army Special Operations Aviation positions at any level
- serving in assignments such as Human Resources Command as the WO branch and career manager or division or brigade staff as an aviation advisor or mission planner.

CW5s are master-level technical and tactical experts that provide support to brigade, division, corps, echelons above corps, and major command level. CW5s have special warrant officer leadership and representation responsibilities within their respective commands. Responsibilities of a CW5 may include

- completing the Warrant Officer Senior Staff Course not later than one year after promotion to CW5
- serving in assignments as the aviation safety officer at the brigade level and above
- serving as the standardization instructor pilot and standardization officer, performing duties above the brigade level

- above the brigade level, serving as the senior special operations aviator; material officer; brigade, division, corps, or Department of the Army–level staff; or chief engineering test pilot
- seeking nominative positions as Command Chief and chief warrant officer of the Army Aviation Branch.

Army Aviation ADSO

Army officers who completed initial entry flight training on or after October 1, 2020, incur a ten-year ADSO. Individuals who entered into a service agreement before October 1, 2020, incur a six-year ADSO (Rempfer, 2020). The service obligation begins on the effective date an officer attains an aeronautical rating of Army aviator or voluntarily terminates attendance.

Table A.2 describes the lengths of different types of training for warrant officer aviators.

Timing of Aviation Career Milestones in TAPDB

In this section, we present graphs of the percentage of commissioned or warrant officers from each entering cohort who complete each milestone (with the career milestones described in the previous section and in Chapter 2) in each YAS to determine what YAS the DRM should use for progression to the next aviation milestone.

Figure A.5 presents the timing of gaining pilot status for commissioned officers and warrant officers. The total bar height represents the percentage of commissioned or warrant officers who entered the Army Aviation Branch during the given calendar year who ever achieved pilot status, whereas the colored portions of each bar represent the percentage of the calendar year's entrants who achieved pilot status during a particular YAS or group of YAS. The graphs suggest that pilot status is typically achieved between 2–3 YAS. Using this analysis, we placed the milestone of achieving pilot status at 2 YAS in the DRM.

The next step in an aviation career is the achievement of pilot-in-command. Figure A.6 presents the timing of achieving pilot-in-command for commissioned officers and warrant officers. The total bar height represents the percentage of commissioned or warrant officers who entered the Army Aviation Branch during the

TABLE A.2
Warrant Officer Aviator Training with a Ten-Year ADSO

Aviator Training	Duration
Direct Commission (ROTC)	n/a
Basic Training	8 weeks
Warrant Officer Candidate School	12 weeks
Officer Candidate School	12 weeks
IERW[a]	32 weeks
Apache Training	64 weeks
Blackhawk/Chinook Training	48 weeks

SOURCE: Features data from Department of the Army Pamphlet 600-3, 2022.

NOTE: n/a = not applicable; ROTC = Reserve Officers' Training Corps.

[a] IERW Training is 32 weeks and consists of four phases. Upon graduation the students will have accumulated 179 hours of flight instruction totaling 149 in an aircraft and 30 in a simulator (Department of the Army Pamphlet 600-3, 2022). Phase 1 consists of two weeks of preflight instruction, providing students with knowledge of basic flight control relationships, aerodynamics, weather, and start-up procedures. Phase 2 (primary phase) consists of ten weeks and 60 flight hours in the TH-67 Creek training helicopter learning the fundamentals of flight, conducting first solo flights, and learning to perform approaches and basic stage field maneuvers. Phase 3 is eight weeks of instrument training, including 30 hours in the flight simulator on the main post and 20 hours in the TH-67. Phase 4 is the combat skills and dual track phase. It is combat mission–oriented and trains the student pilot in the OH-58 A/C as an aeronautical scout helicopter pilot.

FIGURE A.5
Timing of Achievement of Pilot Status for Commissioned Officers and Warrant Officers

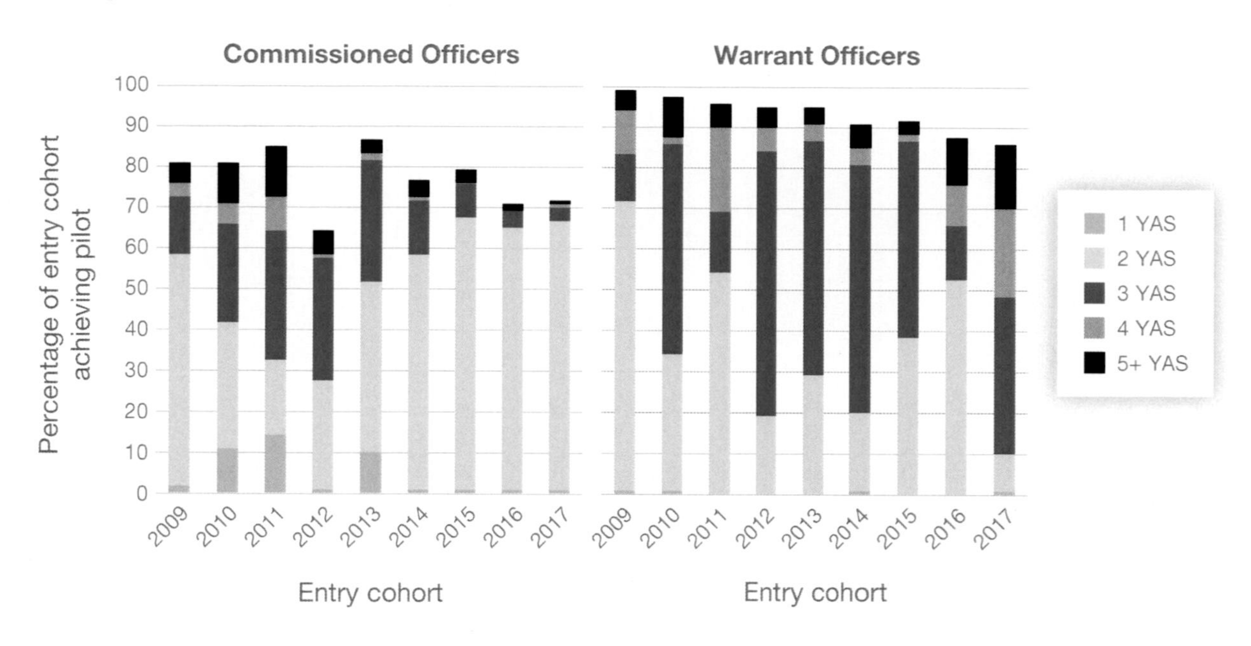

SOURCE: Authors' calculations using TAPDB

FIGURE A.6
Timing of Achievement of Pilot-in-Command for Commissioned and Warrant Officers

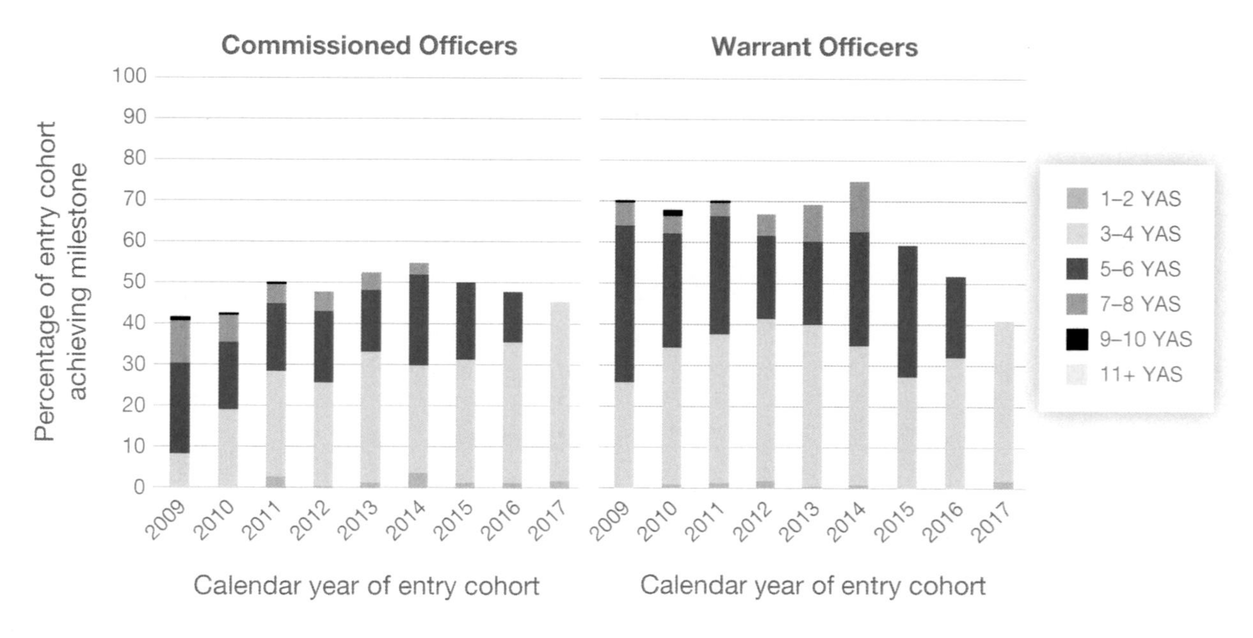

SOURCE: Authors' calculations using TAPDB.

given calendar year who ever achieved pilot-in-command, whereas the colored portions of each bar represent the percentage of the calendar year's entrants who achieved pilot-in-command during a particular YAS or group of YAS. Note that later YAS are censored for later cohorts. The graphs suggest that pilot-in-command is typically achieved between 3–4 YAS. Using this analysis, we placed the milestone of achieving pilot-in-command at 4 YAS in the DRM.

We next examined the first PME of the career, which we did separately for commissioned officers versus warrant officers. Figure A.7 presents the timing of the Captain's Career Course for commissioned officers. The total bar height represents the percentage of commissioned officers who entered the Army Aviation Branch during the given calendar year who ever completed the Captain's Career Course, whereas the colored portions of each bar represent the percentage of the calendar year's entrants who completed the Captain's Career Course during a particular YAS or group of YAS. Note that later YAS are censored for later cohorts. The graphs suggest that the Captain's Career Course is typically completed between 5–6 YAS. Drawing on this analysis, we placed the milestone of completing the Captain's Career Course at 6 YAS in the DRM for commissioned officers. Figure A.8 presents the timing of the warrant officer intermediate course for warrant officers. The total bar height represents the percentage of warrant officers who entered the Army Aviation Branch during the given calendar year who ever completed the warrant officer intermediate course, whereas the colored portions of each bar represent the percentage of the calendar year's entrants who completed the warrant officer intermediate course during a particular YAS or group of YAS. Note that later YAS are censored for later cohorts. The graphs suggest that the warrant officer intermediate course is typically completed between 8–9 YAS, which is considerably later than suggested by discussions with our sponsor's office. We believe that the discrepancy may be explained by either delayed record-keeping or by changes over time in either the typical timing of the course or the way the course is recorded in the TAPDB. Using this analysis, we excluded the milestone of completing the warrant officer intermediate course from the DRM.

The next milestone for warrant officers is tracking. Figure A.9 presents the timing of achieving tracking for warrant officers. The total bar height represents the percentage of warrant officers who entered the Army Aviation Branch during the given calendar year who entered a track prior to September 2021, whereas the col-

FIGURE A.7

Timing of Captain's Career Course Completion for Commissioned Officers, by YAS

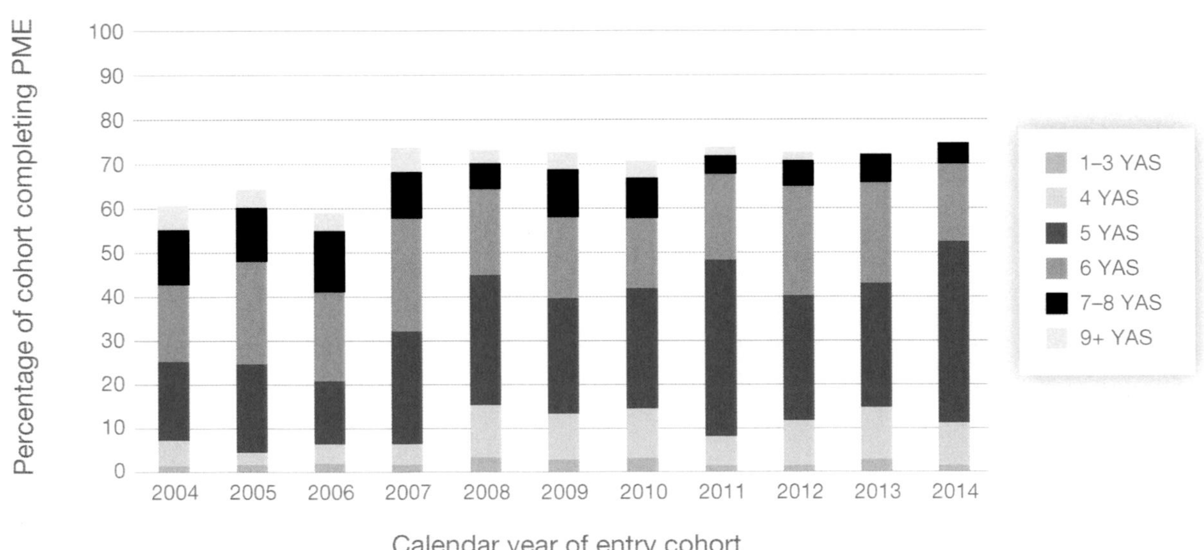

SOURCE: Authors' calculations using TAPDB.

FIGURE A.8
Timing of Warrant Officer Intermediate Course Completion for Warrant Officers

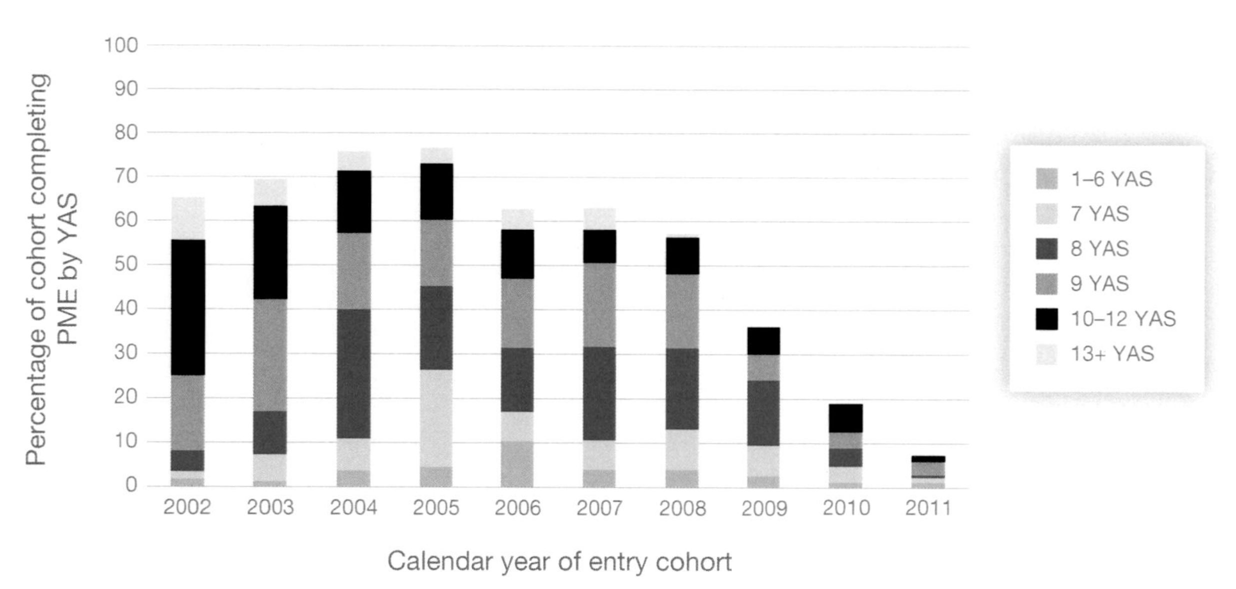

SOURCE: Authors' calculations using TAPDB.

FIGURE A.9
Timing of Tracking for Warrant Officers

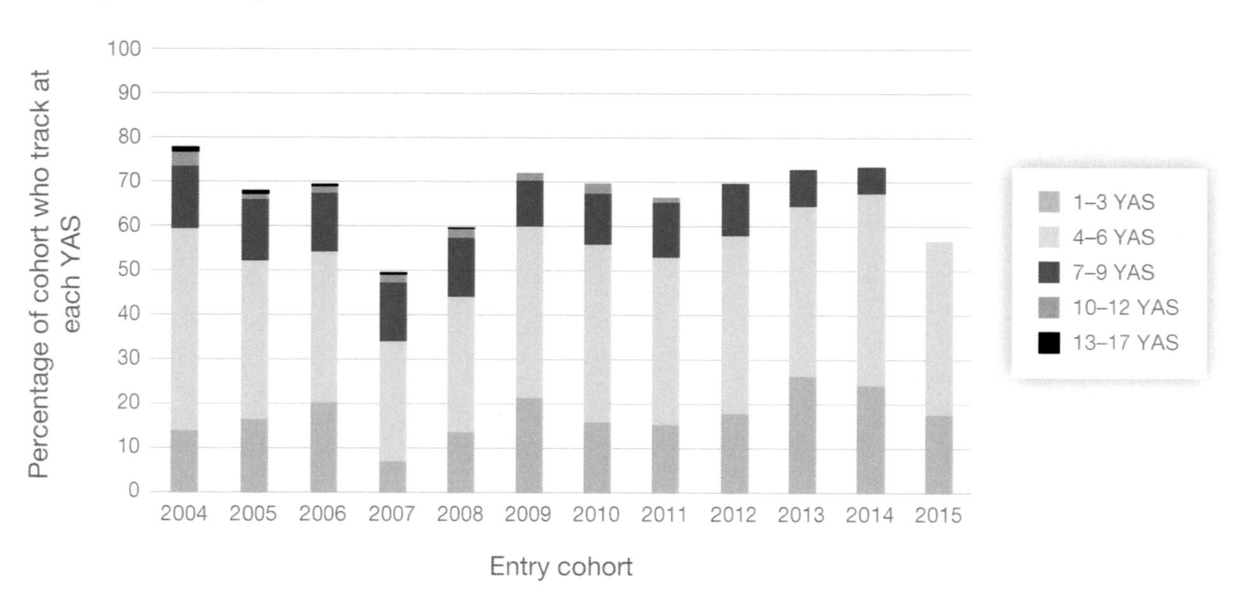

SOURCE: Authors' calculations using TAPDB.

ored portions of each bar represent the percentage of the calendar year's entrants who entered a track during a particular YAS or group of YAS. Note that later YAS are censored for later cohorts. The graphs suggest that tracking is typically achieved between 4–6 YAS. Drawing on this analysis, we placed the milestone of tracking for warrant officers at 6 YAS in the DRM.

The next step in an aviation career is the achievement of senior aviator. Figure A.10 presents the timing of achieving senior aviator for commissioned officers and warrant officers. The total bar height represents the percentage of commissioned or warrant officers who entered the Army Aviation Branch during the given calendar year who ever achieved senior aviator, whereas the colored portions of each bar represent the percentage of the calendar year's entrants who achieved senior aviator during a particular YAS or group of YAS. Note that later YAS are censored for later cohorts. The graphs suggest that senior aviator is typically achieved between 8–9 YAS for commissioned officers and between 10–12 YAS for warrant officers. Using this analysis, we placed the milestone of achieving senior aviator at 8 YAS for commissioned officers in the DRM and at 10 YAS for warrant officers in the DRM.

The final milestone we examined in an aviation career is the achievement of master aviator. Figure A.11 presents the timing of achieving master aviator for commissioned officers and warrant officers. The total bar height represents the percentage of commissioned or warrant officers who entered the Army Aviation Branch during the given calendar year who ever achieved master aviator, whereas the colored portions of each bar represent the percentage of the calendar year's entrants who achieved master aviator during a particular YAS or group of YAS. Note that later YAS are censored for later cohorts, and that because master aviator happens late in the career, we had very few cohorts to work with. The graphs suggest that master aviator is typically achieved between 16–17 YAS. Using this analysis, we placed the milestone of achieving master aviator at 16 YAS in the DRM.

FIGURE A.10
Timing of Senior Aviator Achievement for Commissioned and Warrant Officers

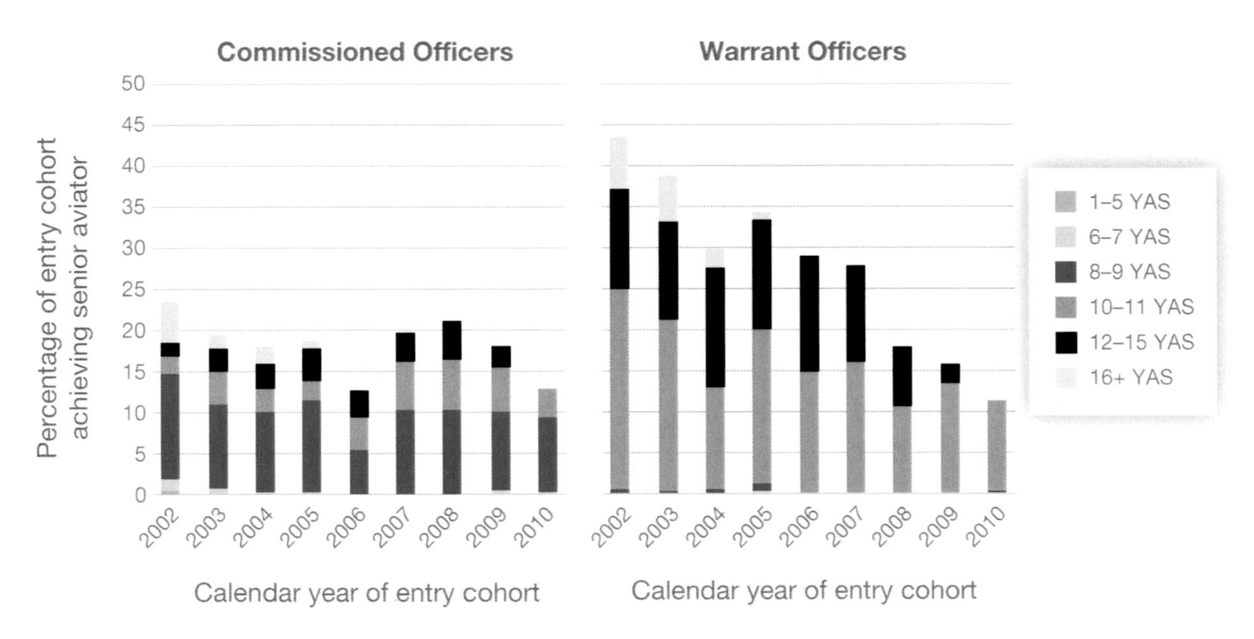

SOURCE: Authors' calculations using TAPDB.

FIGURE A.11
Timing of Master Aviator Achievement for Commissioned and Warrant Officers

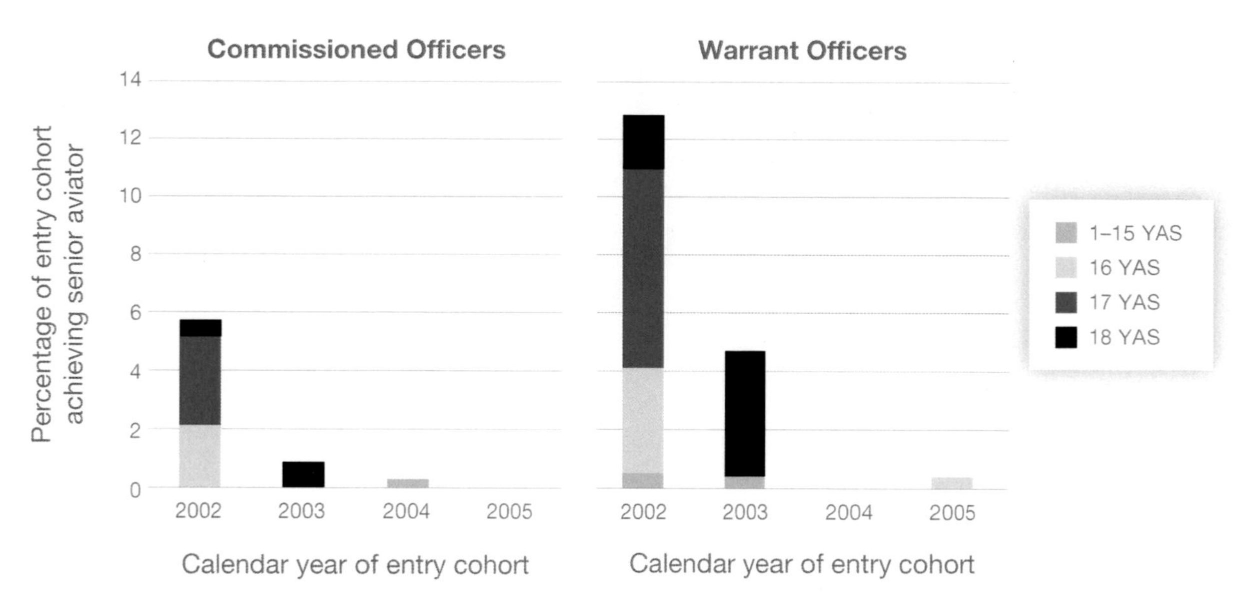

SOURCE: Authors' calculations using TAPDB.

APPENDIX B

Army Aviation Special and Incentive Pays

This appendix provides information on S&I pay levels and qualifications for Army aviators. We begin with a discussion of AvIP and follow with a discussion of AvB. We also include information on the values of AvB incorporated into the DRM.

AvIP Values and Qualification for AvIP

AvIP for Rated Aviation Officers

AvIP begins when aviation service entry date orders are published for select personnel who have completed all required aviation training. Eligibility to receive AvIP is terminated when an aviator fails to meet aviator training certifications or exceeds 25 YAS (except warrant officers) (AR 600-105, 2020).[1] Tables B.1 and B.2 present AvIP rates from FY 1998–July 2020 and rates as of FY 2020.

TABLE B.1
AvIP Rates, FY 1998–July 2020

YAS	Monthly Amount ($)
2 or Less	125
Over 2	156
Over 3	188
Over 4	206
Over 6	650
Over 14	840
Over 22	585
Over 23	495
Over 24	385
Over 25	250

SOURCE: Features data from DoDI 7000.14-R, 2001.

TABLE B.2
AvIP Rates as of FY 2020

YAS	Monthly Amount ($)
2 or Less	125
Over 2	200
Over 6	700
Over 10	1,000
Over 22	700
Over 24	400

SOURCE: Features data from DoDI 7000.14-R, 2001.

[1] Qualified aviators may receive AvIP on a conditional basis provided the requirements in AR 600-105 (2020) are met. Eligibility is terminated when an aviator becomes medically disqualified, terminated because of Flying Evaluation Board, or approved for branch transfer or change in area of concentration.

Qualifications for AvIP

As described in AR 600-105 (2020), officers eligible for career AvIP or conditional AvIP include

- officers entitled to basic pay in the grade of WO1 through colonel who hold an aeronautical rating
- students in flight training
- current Class I or II flight physical
- officers with a pilot status Codes 1 and 5
- officers who have achieved the minimum flight requirements of four hours during one calendar month or 24 hours during six consecutive months.[2]

Army aviators who remain part of the rated inventory are eligible for career AvIP through the first 12 YAS. After 12 YAS, aviators with sufficient total operational flying duty credit, as listed in Table 3–1 of AR 600-103, will remain eligible for career AvIP if they remain assigned to one or more of the following categories, also described in AR 600-105 (2020):

- aviation-specific positions that must be filled by officers with an aeronautical rating
- joint assignments or positions on the Joint Duty Assignment List
- attending resident PME or a fully funded graduate degree program authorized by the Secretary of the Army
- other career-enhancing assignments outside aviation or based on the needs of the Army (voluntary or involuntary) for a period not to exceed 48 consecutive months.

Historical Values of AvB

This section presents a history of AvB. AvB is a retention incentive for Army warrant officer pilots. AvB takes the form of a bonus, paid either yearly or up front as a lump sum, tied to a multi-year ADSO. The eligibility requirements for, targeting, amount, and ADSO length for AvB have varied over our study period.

Table B.3 describes eligibility for and amount of AvB in each FY from 2007 to 2022. In some years, multiple amounts were offered, depending on a variety of conditions as described in the table. The most common contract length was three years, although contracts of different lengths were occasionally offered, most often to special operations aviators. From FY 2007 to 2012, AvB was typically offered to all warrant officers in particular communities, such as special operations aviators, tactical operations, and MTFEs. These communities are typically identified in the data using SQIs. In certain years during this period, the Army also targeted warrant officers with 19 to 24 years of AFS who held SQIs that corresponded to being tracked. AvB was not available between FY 2013 and FY 2017. From FY 2018 on, warrant officers had to be between the ranks of CW2 and CW4, have completed or be within one year of completing their initial ADSO, be eligible for AvIP, and be and remain medically qualified for flight duty to be eligible for AvB, in addition to the requirements noted in the table. Additional eligibility requirements, such as being less than six years time-in-grade or not serving in a special operations billet, occurred in some but not all years after FY 2018. Warrant officers with 14 to 18 years of AFS are currently legally barred from receiving AvB (U.S. Army Human Resources Command, 2022).

[2] The minimum flight hour requirement for reserve component officers not on active duty for a period of more than 30 consecutive days is two hours during one calendar month or 12 hours during six consecutive months. A certified flight simulator may be used to meet this requirement.

TABLE B.3
AvB Eligibility and Levels by Fiscal Year

FY	AvB Offered?	Eligible MOS or Aircraft	Years of Service Eligible	Community, Aircraft, or Required Milestone	Amount of AvB	Contract Length	Source
2007	Yes	152C/D/F/H, 153D/E, 154C/E, 155A/D/E/F/G		Special operations	$12,000–$25,000 per year	2–4 years	MILPER 06-318
		152B/C/D/F/H, 153A/B/D/E, 154C/E		MTPs and MTFEs	$12,000 per year	3 years	MILPER 06-319
		152B/C/D/F/H, 153A/B/D/E, 154C/E		Tactical operations	$12,000 per year	3 years	MILPER 06-319
2008	Yes	152C/D/F/H, 153D/E, 154C/E, 155A/D/E/F/G		Special operations	$12,000–$25,000 per year	2–4 years	MILPER 07-306
		152B/C/D/F/H, 153A/B/D/E, 154C/E		MTPs and MTFEs	$15,000 per year	3 years	MILPER 07-305
		152B/C/D/F/H, 153A/B/D/E, 154C/E		Tactical operations	$12,000 per year	3 years	MILPER 07-307
		152F/H, 154C/F	19 to 24 years AFS	Tracked	$25,000	3 years	MILPER 08-181
2009	Yes	152C/D/F/H, 153D/E, 154C/E, 155A/D/E/F/G		Special operations	$12,000–$25,000 per year	2–4 years	MILPER 08-257
		152B/C/D/F/H, 153A/B/D/E, 154C/E		MTPs and MTFEs	$12,000 per year	3 years	MILPER 08-256
		152B/C/D/F/H, 153A/B/D/E, 154C/E		Tactical operations	$12,000 per year	3 years	MILPER 08-258
		152F/H, 154C/F	19 to 24 years AFS	Tracked	$25,000 per year	3 years	MILPER 08-258
2010	Yes	152C/D/F/H, 153D/E, 154C/E, 155A/D/E/F/G		Special operations	$12,000–$25,000 per year	Over 1 to 6 years	MILPER 09-246
2011	Yes	152C/D/F/H, 153D/E, 154C/E, 155A/D/E/F/G		Special operations	$12,000–$25,000 per year	Over 1 to 6 years	MILPER 11-053
		AH-64A/D (aircraft) or 152F/H (MOS)	19 to 24 years AFS	Instructor pilots, instrument flight examiners, standardization instructor pilots	$25,000 per year	3 years	MILPER 11-054
2012	Yes	152C/D/F/H, 153D/E, 154C/E, 155A/D/E/F/G		Special operations	$12,000–$25,000 per year	Over 1 to 6 years	MILPER 11-054
2013	No						

Table B.3—Continued

FY	AvB Offered?	Eligible MOS or Aircraft	Years of Service Eligible	Community, Aircraft, or Required Milestone	Amount of AvB	Contract Length	Source
2014	No						
2015	No						
2016	No						
2017	No						
2018	Yes	152H/E, 153A/D/M, 154F, 155A/E/F/G	Less than 11 years AFS	Pilot-in-command	152H/E: $15,000 per year; 153A/D/M, 154F, 155A/E/F/G: $12,000 per year	At least 3 years, until 14 years AFS but not past 15 years AFS	ALARACT 085/2017
				Track	152H/E: $18,000 per year; 153A/D/M, 154F, 155A/E/F/G: $15,000 per year		
			19 to 22 years AFS	Pilot-in-command	152H/E: $18,000 per year; 153A/D/M, 154F, 155A/E/F/G: $15,000 per year	At least 3 years, until 24 years AFS but not past 25 years AFS	
				Track	152H/E: $21,000 per year; 153A/D/M, 154F, 155A/E/F/G: $18,000 per year		
2019	Part year only (October–December 2018)	152H/E, 153D/M, 154F, 155E/F/G	No more than 13 years AFS	Pilot-in-command	152H/E: $45,000 over 3 years	3 years	ALARACT 071/2018
				Track	152H/E: $60,000 over 3 years 153D/M, 154F: $45,000 over 3 years		
			19 to 22 AFS	Pilot-in-command	152H/E: $55,000 over 3 years		
				Track	152H/E: $90,000 over 3 years 153D/M, 154F: $60,000 over 3 years		
				Fixed-wing aircraft	155E/F/G: $105,000 over 3 years		
2020	Yes		No more than 13 years AFS	Tracked	$30,000 per year	3 years	ALARACT 075/2019
				Fixed-wing aircraft	$30,000 per year		
			19 to 22 years AFS	Tracked	$30,000 per year		
				Fixed-wing aircraft	$30,000 per year		
2021	No						

NOTE: MILPER = Military Personnel (Message).

We simplified the schedule of available AvB when including AvB in the DRM because our data did not break down warrant officers by MOS and YOS and because warrant officers may shift across MOSs over time. Typically, we used the amount of AvB per year that corresponded to the largest number of individuals in our data or was the average of available amounts, removed AvB offers that applied to less than one-half of the fiscal year, assumed all pilots flew rotary-wing aircraft, assumed no pilots were assigned to special operations billets, and treated all SQI as if those warrant officers were tracked. We assumed that the current restriction on receiving AvB for warrant officers with 14 to 18 years AFS also applied from FY 2007 to 2012, as well as the requirement to have completed the initial ADSO of six years. We excluded AvB targeting retirement-eligible personnel from FYs 2008, 2009, and 2011 because no one in our sample was eligible for those AvBs. Finally, we assumed that all contracts were three years long. Table B.4 reports the AvB information included in the DRM.

TABLE B.4

AvB Eligibility and Levels Included in the DRM

FY	AvB Offered?	YOS Eligible	Required Milestone	Amount of AvB	Contract Length
2007	Yes	7–13 years AFS	Tracked	$12,000 per year	3 years
2008	Yes	7–13 years AFS	Tracked	$12,000 per year	3 years
2009	Yes	7–13 years AFS	Tracked	$12,000 per year	3 years
2010	No				
2011	No				
2012	No				
2013	No				
2014	No				
2015	No				
2016	No				
2017	No				
2018	Yes	7–11 years AFS	Pilot-in-command	$15,000 per year	3 years
			Track	$18,000 per year	3 years
		19–22 years AFS	Pilot-in-command	$18,000 per year	3 years
			Track	$21,000 per year	3 years
2019	No				
2020	Yes	7–13 years AFS	Tracked	$30,000 per year	3 years
		19–22 AFS	Tracked	$30,000 per year	3 years
2021	No				
2022 and later	Yes	7–13 or 19–22 years AFS	Tracked	$30,000 per year	3 years

SOURCE: Features data from sources noted in Table B.3.

APPENDIX C

Civilian Earnings Analysis

This appendix, which draws heavily on text from Mattock et al. (2016), describes the process we used to calculate percentiles of civilian earnings in the ACS.

ACS Sample

We constructed civilian earnings for veterans using ACS data from 2003 to 2019 (Ruggles et al., 2022). Earnings are calculated using the variable for "wage and salary income" and are converted to 2019 dollars using the CPI-U, constructed by the U.S. Bureau of Labor Statistics (undated). Because the ACS collects information on income from the prior year, the year for earnings is the year before the survey year. We therefore had earnings data for 2002 to 2018. We used only full-time (at least 40 hours per week), full-year (50–52 weeks per year) civilian workers ages 18 to 65 with non-zero earnings and at least some college for our analysis. The earnings estimates should therefore reflect expected earnings from full-time employment. The education requirement was chosen to reflect the typical level of education of Army pilots, who are typically either warrant officers or commissioned officers. Warrant officers most often have some college education (but may have only a high school education), whereas officers almost always have at least four years of college education.

ACS data on income is *top-coded*, meaning that incomes above a certain threshold (calculated separately for each state) are not reported to protect the privacy of high-earning respondents. Individuals whose income is above that threshold have income imputed to be equal to the mean income in their state of residence among individuals whose income is higher than the threshold. This was problematic for our analysis because we used only a portion of the ACS sample, and the mean income for imputed individuals in our subsample may not match that of imputed individuals across the ACS sample, potentially creating statistical bias (which could go in either direction). We therefore used a right-censored *Tobit model* to estimate our earnings regressions. The Tobit model takes the top-coding (or right-censoring, because income levels on the far right of the distribution are not observed) into account by using a two-part equation for the sample likelihood. Censored observations are given a likelihood value equal to the probability of falling in the censored portion of the distribution, whereas non-censored observations are given a likelihood value equal to the probability of earning that exact earnings value. Dropping these observations as missing would likely result in a downward bias of the regression estimates (causing the estimated earnings percentiles to grow too slowly with age).

The sample means for key variables by veteran status are shown in Table C.1. Veterans earn more than non-veterans. This likely partly reflects earnings growth with age combined with the fact that veterans tend to be older than non-veterans. Veterans are less likely than non-veterans to be female: Only 12 percent of veterans are female, as opposed to 47 percent of non-veterans. Conditional on having completed at least one year of college, veterans also have different education levels than non-veterans: 53 percent of veterans have one to three years of college education, 29 percent have four years of college education, and 18 percent have more than four years of college education, as opposed to 40, 39, and 22 percent, respectively, of non-veterans. Finally, 26 percent of veterans in the ACS subsample served in the military in 2001 or later.

TABLE C.1
Summary Statistics by Veteran Status

Statistic Description	Veterans	Non-Veterans
Earnings (2019 dollars)	$85,271.40 ($69,515.10)	$80,767.40 ($76,579.60)
Age (years)	47.58 (10.79)	41.48 (11.36)
Female (%)	12 (32)	47 (50)
1 to 3 years of college (%)	53 (50)	40 (49)
Exactly 4 years of college (%)	29 (45)	39 (49)
More than 4 years of college (%)	18 (38)	22 (41)
Served 2001 or later (%)	27 (44)	
Observations (*n*)	604,507	7,246,741

SOURCE: Authors' calculations using ACS data (U.S. Bureau of Labor Statistics, undated).

NOTE: Summary statistics only include individuals who work full-time all year and who have at least one year of college education. Standard deviations are in parentheses.

Table C.1 omits the means of the year indicators. Between 6 and 6.5 percent of the sample falls into each survey year from 2005–2019, with smaller groups (about 2.5 percent each) falling into the 2003 and 2004 survey years. The table also omits the means for the threshold quartile indicators. The threshold quartile indicators were created so that one-quarter of state-year observations fell into each threshold after rescaling the thresholds to 2019 dollars. However, because of differences in state populations, individuals are not spread evenly across states. The threshold indicators were included to guard against the possibility that older, higher-earning pilots are concentrated in higher-earning states, which could bias the coefficients of the earnings regressions.

Tobit Regression Specifications

We first converted earnings to a log scale, and we assumed that the error term in earnings is normally distributed. We ran our regressions separately for pilots and non-pilots, giving us a total of two regressions. We used the following specification:

$$\ln(\text{Earnings}_i) = \beta_0 + \beta_1 \text{Age}_i + \beta_2 \text{Age}_i^2 + \beta_3 \text{Female}_i + \beta_4 \text{FourYearsColl}_i + \beta_5 \text{MoreThanFourYearsColl}_i + \beta_6 \text{Veteran}_i + \beta_7 \text{Veteran01}_i + \sum_{t=2004}^{2019} \theta_t + \sum_{q=2}^{4} \theta_q + \varepsilon_i.$$

Here, Age_i is a numeric variable for age, Female_i is an indicator variable that *i* is female, FourYearsColl_i is an indicator variable that *i* has exactly four years of college education, $\text{MoreThanFourYearsColl}_i$ is an indicator variable that *i* has more than four years of college education, Veteran_i is an indicator variable that *i* is a veteran, Veteran01_i is an indicator variable that *i* is a veteran who served in 2001 or later, θ_t is an indicator variable for survey year t, and θ_q is a binary indicator that i lives in a state where the top-code threshold for earnings is in quartile q. As is standard for the Tobit model, we assumed

$$\varepsilon \sim \mathcal{N}(0, \sigma^{tobit2}).$$

The Tobit model returns estimates for each of the coefficients on the right-hand side of the regressions, as well as the variance of the error term, using maximum likelihood estimation.

Tobit Regression Estimates

Table C.2 reports the Tobit regression results. We found that earnings increase more rapidly with age for pilots than for non-pilots (similar to the findings of Mattock et al., 2016). This is reflected in our estimates of earnings percentiles for pilots and non-pilots.

Predicted Earnings for Veterans by Earnings Percentile, Age, and Education

Following Mattock et al. (2016), we predicted earnings by sample, age, level of education, and earnings percentile. The predictions were for male veterans who served on active duty in the military in 2001 or later. The regressions included year effects, and we used the year 2018 (from survey year 2019) in the predictions. All predicted earnings are in 2019 dollars. We used the earnings percentiles for individuals with one to three years of college for the warrant officer analyses and for individuals with four years of college for the commissioned officer analysis. We predicted earnings percentiles for individuals between the ages of 22 and 65.

We calculated the 40th, 50th, 60th, 70th, 80th, and 90th percentile of earnings for pilots and non-pilots for each age a, education level j, and earnings quartile q, setting Female_i equal to 0, Veteran_i and Veteran01_i equal to 1, and t equal to 2018. Each percentile of earnings ϑ is then calculated using

$$\ln(\text{Earnings})_q^{\vartheta} = X\hat{\beta} + \Phi(\vartheta/\sigma^{tobit})$$

where ϑ is between 0 and 1 (e.g., the 80th percentile has $\vartheta = 0.8$), $\hat{\beta}$ is the estimated Tobit coefficients from Equation 1, $\Phi(\cdot)$ is the normal cumulative distribution function, and σ^{tobit} is the square root of the variance estimate from Equation 1. We then converted the earnings percentiles from logs to levels, so that we had percentiles of real earnings rather than of log earnings. Finally, the percentiles of earnings were aggregated across the four quartiles of earnings thresholds using

$$\text{Earnings}^{\vartheta} = \sum_{q=1}^{4} \omega^{\vartheta} \text{Earnings}_q^{\vartheta}$$

where $\text{Earnings}_q^{\vartheta}$ is percentile ϑ of earnings in quartile q and ω^{ϑ} is a weight representing the fraction of pilots or non-pilots, whichever we were calculating percentiles for, who live in each state earnings quartile.

Tables C.3 and C.4 present the earnings quartiles for individuals with one to three years of college and for individuals with four years of college, respectively.

TABLE C.2

Right-Censored Tobit Model Regression Estimates for Ln(Earnings)

Variable	Coefficient	Standard Error
Age	0.0882***	0.000182
Age squared	−0.000868***	2.14e−06
Female indicator	−0.273***	0.000548
Education = 4 Years College	0.347***	0.000610
Education = More than 4 Years College	0.587***	0.000734
Veteran indicator	−0.0129***	0.00123
Veteran 2001 or later indicator	0.0246***	0.00212
2004 Year Effect	0.154***	0.00245
2005 Year Effect	0.144***	0.00243
2006 Year Effect	−0.0222***	0.00215
2007 Year Effect	−0.0604***	0.00213
2008 Year Effect	−0.0746***	0.00213
2009 Year Effect	−0.0630***	0.00213
2010 Year Effect	−0.00207	0.00219
2011 Year Effect	−0.0351***	0.00215
2012 Year Effect	−0.0529***	0.00215
2013 Year Effect	−0.0791***	0.00215
2014 Year Effect	−0.0607***	0.00214
2015 Year Effect	−0.0606***	0.00214
2016 Year Effect	−0.0456***	0.00214
2017 Year Effect	−0.0651***	0.00213
2018 Year Effect	−0.0644***	0.00215
Quartile of real top-code threshold for earnings = 2	0.0703***	0.00136
Quartile of real top-code threshold for earnings = 3	0.117***	0.00134
Quartile of real top-code threshold for earnings = 4	0.245***	0.00131
Variance	0.355***	0.000579
Constant	8.718***	0.00428
Observations	7,851,248	
Censored	276,376	

NOTE: Standard errors are robust. *** $p < 0.01$, ** $p < 0.05$, * $p < 0.1$.

TABLE C.3

Earnings Percentiles for Veterans with One to Three Years of College

Age	40th Percentile of Earnings ($)	50th Percentile of Earnings ($)	60th Percentile of Earnings ($)	70th Percentile of Earnings ($)	80th Percentile of Earnings ($)	90th Percentile of Earnings ($)
22	26,155	30,415	35,370	41,568	50,215	65,260
23	27,472	31,948	37,152	43,662	52,745	68,548
24	28,806	33,499	38,956	45,783	55,306	71,876
25	30,153	35,065	40,777	47,922	57,891	75,236
26	31,507	36,640	42,609	50,075	60,491	78,616
27	32,865	38,219	44,446	52,234	63,099	82,005
28	34,223	39,798	46,281	54,392	65,705	85,392
29	35,575	41,370	48,109	56,540	68,301	88,765
30	36,916	42,930	49,923	58,671	70,875	92,111
31	38,241	44,471	51,715	60,778	73,420	95,418
32	39,545	45,987	53,479	62,850	75,924	98,672
33	40,823	47,473	55,207	64,881	78,377	101,860
34	42,069	48,922	56,892	66,861	80,769	104,969
35	43,277	50,328	58,526	68,782	83,089	107,985
36	44,444	51,684	60,103	70,636	85,329	110,895
37	45,562	52,985	61,616	72,414	87,476	113,686
38	46,628	54,224	63,058	74,108	89,523	116,345
39	47,636	55,396	64,421	75,710	91,458	118,861
40	48,582	56,496	65,699	77,212	93,273	121,220
41	49,460	57,517	66,887	78,608	94,959	123,411
42	50,267	58,456	67,978	79,891	96,509	125,425
43	50,998	59,306	68,968	81,053	97,913	127,250
44	51,651	60,065	69,850	82,090	99,166	128,878
45	52,221	60,728	70,621	82,996	100,260	130,300
46	52,706	61,292	71,277	83,767	101,191	131,510
47	53,103	61,754	71,814	84,398	101,954	132,501
48	53,410	62,111	72,229	84,887	102,544	133,268
49	53,626	62,362	72,522	85,230	102,959	133,807
50	53,750	62,506	72,689	85,426	103,196	134,115
51	53,780	62,542	72,730	85,475	103,254	134,191
52	53,718	62,468	72,645	85,375	103,134	134,035
53	53,562	62,287	72,434	85,128	102,835	133,646
54	53,314	61,999	72,099	84,734	102,359	133,028

Table C.3—Continued

Age	40th Percentile of Earnings ($)	50th Percentile of Earnings ($)	60th Percentile of Earnings ($)	70th Percentile of Earnings ($)	80th Percentile of Earnings ($)	90th Percentile of Earnings ($)
55	52,975	61,605	71,641	84,195	101,708	132,182
56	52,547	61,108	71,062	83,515	100,887	131,115
57	52,033	60,509	70,366	82,697	99,899	129,831
58	51,434	59,812	69,556	81,745	98,749	128,336
59	50,753	59,021	68,636	80,664	97,443	126,638
60	49,995	58,140	67,611	79,459	95,987	124,747
61	49,163	57,172	66,486	78,136	94,389	122,671
62	48,261	56,123	65,266	76,703	92,658	120,420
63	47,293	54,998	63,957	75,165	90,800	118,005
64	46,265	53,801	62,566	73,530	88,825	115,439
65	45,180	52,540	61,099	71,806	86,742	112,732

TABLE C.4

Earnings Percentiles for Veterans with Four Years of College

Age	40th Percentile of Earnings ($)	50th Percentile of Earnings ($)	60th Percentile of Earnings ($)	70th Percentile of Earnings ($)	80th Percentile of Earnings ($)	90th Percentile of Earnings ($)
22	37,000	43,028	50,037	58,806	71,038	92,322
23	38,864	45,195	52,558	61,768	74,616	96,973
24	40,751	47,390	55,110	64,767	78,240	101,682
25	42,656	49,605	57,686	67,795	81,896	106,434
26	44,572	51,833	60,277	70,840	85,576	111,216
27	46,494	54,068	62,876	73,894	89,265	116,011
28	48,414	56,301	65,473	76,946	92,952	120,802
29	50,327	58,525	68,059	79,986	96,623	125,574
30	52,224	60,731	70,625	83,001	100,266	130,308
31	54,099	62,912	73,160	85,981	103,865	134,985
32	55,943	65,057	75,655	88,913	107,407	139,589
33	57,751	67,159	78,099	91,785	110,877	144,099
34	59,513	69,209	80,483	94,587	114,261	148,497
35	61,224	71,197	82,796	97,305	117,545	152,763
36	62,873	73,116	85,027	99,927	120,712	156,880
37	64,456	74,956	87,167	102,442	123,751	160,829
38	65,964	76,710	89,206	104,838	126,646	164,591
39	67,390	78,368	91,135	107,105	129,383	168,149
40	68,727	79,923	92,943	109,230	131,951	171,487
41	69,970	81,368	94,624	111,205	134,337	174,587
42	71,111	82,696	96,167	113,020	136,528	177,435
43	72,146	83,899	97,567	114,664	138,515	180,017
44	73,069	84,973	98,815	116,131	140,287	182,320

Table C.4—Continued

Age	40th Percentile of Earnings ($)	50th Percentile of Earnings ($)	60th Percentile of Earnings ($)	70th Percentile of Earnings ($)	80th Percentile of Earnings ($)	90th Percentile of Earnings ($)
45	73,876	85,910	99,906	117,413	141,836	184,333
46	74,562	86,708	100,833	118,503	143,153	186,044
47	75,123	87,361	101,593	119,396	144,231	187,446
48	75,558	87,867	102,181	120,087	145,066	188,531
49	75,864	88,223	102,595	120,573	145,653	189,294
50	76,039	88,426	102,831	120,851	145,989	189,730
51	76,082	88,476	102,889	120,919	146,071	189,837
52	75,993	88,373	102,769	120,778	145,901	189,616
53	75,773	88,116	102,471	120,428	145,478	189,066
54	75,422	87,709	101,997	119,871	144,805	188,191
55	74,943	87,151	101,349	119,109	143,884	186,995
56	74,337	86,447	100,530	118,147	142,722	185,485
57	73,609	85,601	99,546	116,990	141,324	183,668
58	72,762	84,615	98,399	115,643	139,697	181,554
59	71,800	83,496	97,098	114,113	137,850	179,152
60	70,727	82,249	95,648	112,409	135,791	176,476
61	69,550	80,880	94,056	110,538	133,531	173,539
62	68,274	79,396	92,330	108,510	131,080	170,355
63	66,905	77,804	90,479	106,334	128,452	166,939
64	65,450	76,112	88,511	104,021	125,658	163,308
65	63,915	74,327	86,435	101,582	122,712	159,479

APPENDIX D

Simulation Results with a Six-Year ADSO

As noted in Chapters 5 and 7, our model was estimated using a six-year initial ADSO, but our main simulations were calculated using a ten-year ADSO because of a policy change in FY 2021. Our simulations hold constant the composition of entering aviators relative to the estimated model, which may not be realistic when the initial ADSO increases by four years. Therefore, in this appendix, we present our main simulation results (AvIP simulations without ability and AvB) assuming a six-year initial ADSO.

Figure D.1 repeats the analysis presented in Figure 5.1 under a six-year ADSO. Figure D.2 repeats the analysis presented in Figure 5.2 under a six-year initial ADSO. Table D.1 repeats the analysis presented in Table 5.2 under a six-year initial ADSO. Table D.2 repeats the analysis presented in Table 5.3 under a six-year initial ADSO. Table D.3 repeats the analysis presented in Table 6.1 under a six-year ADSO. Table D.4 repeats the analysis presented in Table 6.2 under a six-year ADSO. Table D.5 repeats the analysis presented in Table 6.3 under a six-year ADSO. Table D.6 repeats the analysis presented in Table 6.4 under a six-year ADSO. Table D.7 repeats the analysis presented in Table 6.5 under a six-year ADSO.

FIGURE D.1
Retention If AvIP Maintained 1998 Real Value with Six-Year ADSO

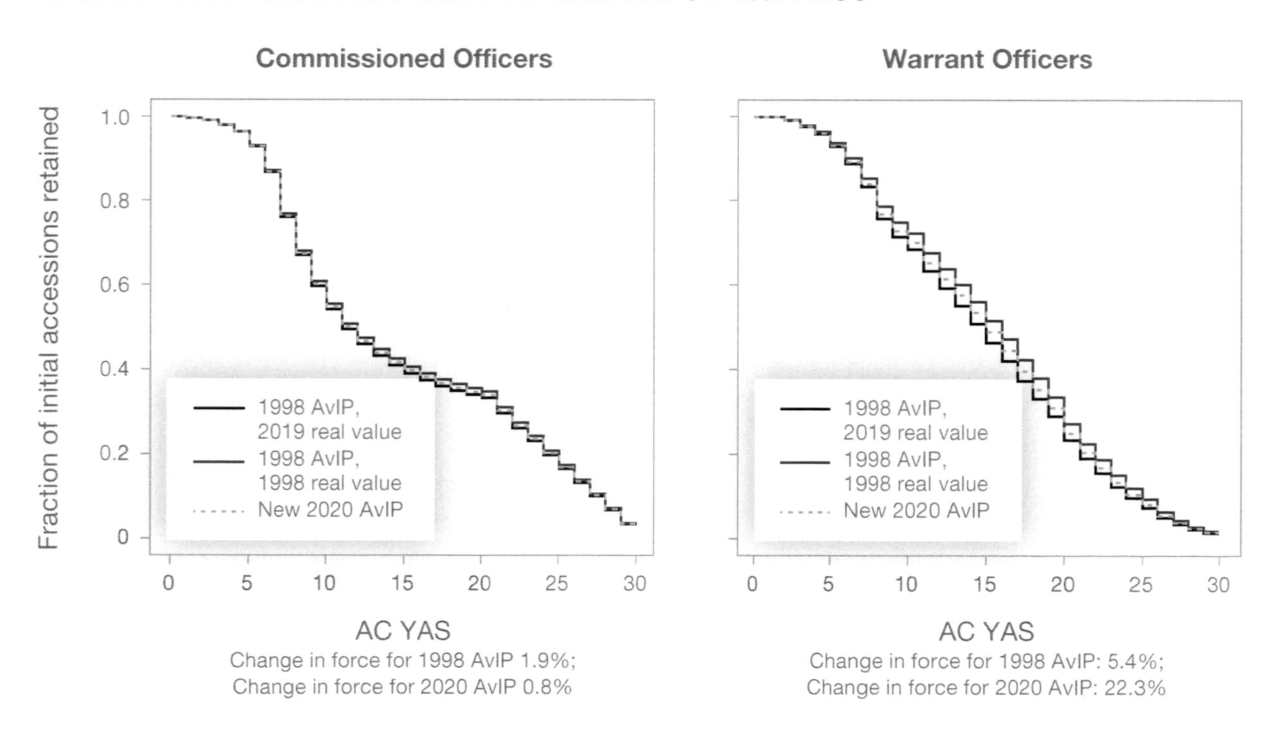

NOTE: Figures report retention of commissioned officers (left) and warrant officers (right) in the Army Aviation Branch under various levels of AvIP. Baseline (black line) refers to the value of AvIP in force from FY 1998 to July 2020 (see Table 2.2). Real value in 1998 (red line) was calculated using the CPI-U Inflator, that is, AvIP * CPI-U (2019)/CPI-U (1998). New 2020 AvIP (blue dashed line) refers to the value of AvIP in force starting January 2020 (see Table 2.3). Simulations assumed a six-year initial ADSO and that tracked warrant officers with 7 to 13 or 19 to 22 years of AFS are eligible for $30,000 per year in AvB over a three-year contract.

FIGURE D.2
Retention Effects of the Army's Prototype AvIP Values Under Six-Year ADSO

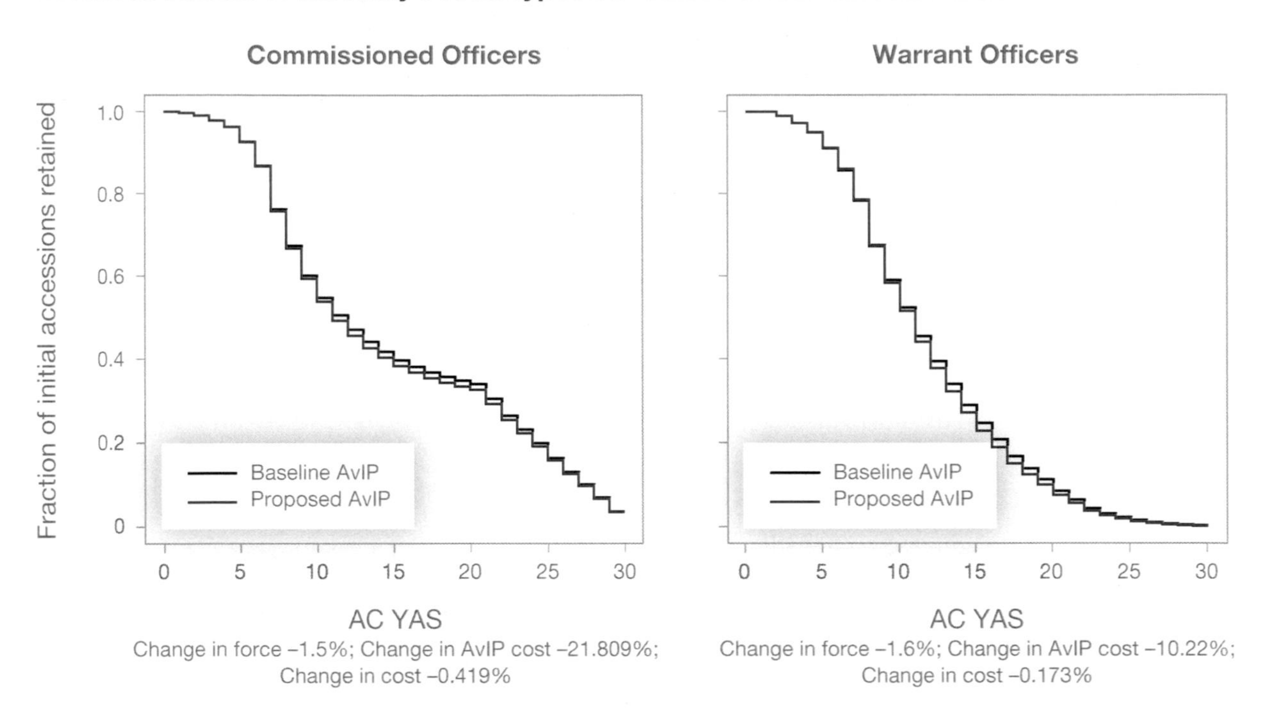

NOTE: Figures show simulations of retention of commissioned officers (left) and warrant officers (right) in the Army Aviation Branch under different ways of setting AvIP. The black line shows retention under the AvIP values set in January 2020 (see Table 2.3). The red line shows retention under the Army's prototype milestone-based AvIP (see Table 5.1). Simulations assumed a six-year initial ADSO and that AvB is not offered to warrant officers.

TABLE D.1
Retention-Maintaining Values of AvIP by Milestone Under Six-Year ADSO

	Commissioned Officers		
AvIP Value by Milestone	Retention-Maintaining AvIP	Retention-Maintaining Without Perverse Incentive	Warrant Officers
Flight school entry	$125	$125	$125
Pilot status	$250	$250	$185
Pilot-in-command	$500	$500	$370
Captain's Career Course (CO) or Track (WO)	$970	$970	$685
Senior aviator	$1,580	$1,580	$1,590
Master aviator	$1,500[a]	$1,580[a]	$1,620
% change in Force Size	0.001%	0.013%	0.003%
% change in AvIP Cost	13.220%	13.490%	6.440%
% change in Overall Cost	0.250%	0.260%	0.100%

NOTE: Table reports results of simulations of retention of commissioned officers in the Army Aviation Branch under different ways of setting AvIP, with AvIP values for each milestone reported in the columns. Baseline retention uses the AvIP values set in January 2020 (see Table 2.3). Simulations assumed a six-year initial ADSO and that AvB is not offered to warrant officers. CO = commissioned officer; WO = warrant officer.

[a] AvIP is $0 for commissioned officers with over 25 YAS.

TABLE D.2

Retention-Maintaining Values of AvIP with Higher Probability of Milestone Achievement

AvIP Value by Milestone	Commissioned Officers (with P [pilot-in-command] = 90%)	Warrant Officers (with P [track] = 90%)
Flight school entry	$125	$125
Pilot status	$150	$175
Pilot-in-command	$350	$230
Captain's Career Course (commissioned officers) or track (warrant officers)	$795	$665
Senior aviator	$1,180	$1,295
Master aviator	$1,190	$1,325
% change in Force Size	−0.001%	−0.002%
% change in AvIP Cost	6.760%	1.966%
% change in Overall Cost	0.130%	0.000%

NOTE: Table reports results of simulations of retention of commissioned officers in the Army Aviation Branch under different ways of setting AvIP, with AvIP values for each milestone reported in the columns. Baseline retention uses the AvIP values set in January 2020 (see Table 2.3). Simulations assumed a six-year initial ADSO and that AvB is not offered to warrant officers.

TABLE D.3

Effect of Different Values of AvB on Warrant Officer Retention Relative to Baseline of No AvB, Under Six-Year ADSO

AvB Amount, Pilots-in-Command ($)	AvB Amount, Tracked Aviators ($)	Overall Effect of AvB on Retention (%)
0	5,000	7.6
0	10,000	9.1
0	15,000	10.7
0	20,000	12.6
0	25,000	14.8
0	30,000	16.8
0	35,000	19.2
0	40,000	21.4
0	45,000	23.1
0	50,000	25.4

NOTE: AvB amount is yearly and assumed a three-year contract. Simulated warrant officers are eligible for AvB beginning after the end of the six-year initial ADSO until 13 years of AFS (including any enlisted service prior to entering the Army Aviation Branch) and from 19 to 22 years of AFS. All simulations assumed the AvIP values in effect beginning January 2020 and included warrant officers both with and without prior enlisted service (with years of enlisted service matching the distribution in the observed sample).

TABLE D.4

Effect of Changes in Unemployment Rate on Warrant Officer Retention With and Without AvB, Under Six-Year ADSO, Relative to 4.1 Percent Unemployment and No AvB at Baseline

Unemployment Rate (%)	Change in Overall Force Size When No AvB Offered (%)	Yearly AvB Value That Maintains Retention (Unemployment Rate Decreases Only) ($)
0	−3.2	0
1	−2.4	0
2	−1.6	0
3	−0.9	0
4	−0.1	NA
5	0.7	NA
6	1.6	NA
7	2.2	NA
8	2.9	NA
9	3.8	NA
10	4.7	NA

NOTE: AvB amount is yearly and assumes a three-year contract. Simulated warrant officers are eligible for AvB beginning after the end of the six -year initial ADSO until 13 years of AFS (including any enlisted service prior to entering the Army Aviation Branch) and from 19 to 22 years of AFS. All simulations assume the AvIP values in effect beginning January 2020 and include warrant officers both with and without prior enlisted service (with years of enlisted service matching the distribution in the observed sample). NA = not applicable.

TABLE D.5

Effect of Changes in Civilian Pay (Relative to Military Pay) on Warrant Officer Retention With and Without AvB, Under Six-Year ADSO, Relative to No Change in Civilian Pay and No AvB

Increase in Civilian Pay Relative to Military Pay (%)	Change in Overall Force Size When No AvB Offered (%)	Yearly AvB Value That Maintains Retention ($)
1	-0.8	0
2	-1.6	0
3	-2.3	0
4	-3.1	0
5	-3.7	0
6	-4.5	0
7	-5.1	2,000
8	-5.9	5,000
9	-6.6	7,000
10	-7.3	11,000

NOTE: AvB amount is yearly and assumed a three-year contract. Simulated warrant officers are eligible for AvB beginning after the end of the six-year initial ADSO until 13 years of AFS (including any enlisted service prior to entering the Army Aviation Branch) and from 19 to 22 years of AFS. All simulations assumed the AvIP values in effect beginning January 2020 and included warrant officers both with and without prior enlisted service (with years of enlisted service matching the distribution in the observed sample).

TABLE D.6

Effect of Changes in Unemployment Rate on Warrant Officer Retention by AvB Level, Relative to a Baseline of $30,000 AvB for Tracked Aviators and 4.1 Percent Unemployment, Six-Year ADSO

Unemployment Rate (%)	Change in Overall Force Size When Tracked Offered $30k AvB (%)	Yearly AvB Value That Maintains Retention (Unemployment Rate Decreases Only) ($)
0	–4	39,000
1	–3	37,000
2	–2	35,000
3	–1	33,000
4	0	NA
5	1	NA
6	2	NA
7	3	NA
8	4	NA
9	5	NA
10	6	NA

NOTE: AvB amount is yearly and assumed a three-year contract. Simulated warrant officers are eligible for AvB beginning after the end of the ten-year initial ADSO until 13 years of AFS (including any enlisted service prior to entering the Army Aviation Branch) and from 19 to 22 years of AFS. All simulations assumed the AvIP values in effect beginning January 2020 and included warrant officers both with and without prior enlisted service (with years of enlisted service matching the distribution in the observed sample). k = thousand.

TABLE D.7

Effect of Changes in Civilian Pay (Relative to Military Pay) on Warrant Officer Retention With and Without AvB Increase, Relative to $30,000 AvB for Tracked Aviators, Six-Year ADSO

Increase in Civilian Pay Relative to Military Pay (%)	Change in Overall Force Size When Tracked Offered $30k AvB (%)	Yearly AvB Value That Maintains Retention ($)
1	–0.9	33,000
2	–1.8	34,000
3	–2.7	37,000
4	–3.6	39,000
5	–4.5	41,000
6	–5.3	43,000
7	–6.1	45,000
8	–6.9	47,000
9	–7.6	49,000
10	–8.2	51,000

NOTE: AvB amount is yearly and assumed a three-year contract. Simulated warrant officers are eligible for AvB beginning after the end of the ten-year initial ADSO until 13 years of AFS (including any enlisted service prior to entering the Army Aviation Branch) and from 19 to 22 years of AFS. All simulations assumed the AvIP values in effect beginning January 2020 and included warrant officers both with and without prior enlisted service (with years of enlisted service matching the distribution in the observed sample).

Estimates Including Major Airline Hiring

This appendix presents DRM estimates when we included the possibility of being hired at a major airline into the value of leaving the Army for a civilian career.

Incorporating Major Airline Hiring into the DRM

We incorporated pilot pay and the probability of being hired at a major airline into the DRM following the methods shown in Mattock et al. (2016). We redefined the present discounted value of all future civilian compensation to be

$$V_t^C = \sum_{s=t}^{T} \beta^{s-1}\left(p(\text{hired})\times W_t^{C,\text{pilot}} + \left(1 - p(\text{hired})\right)\times W_{t-t\_leave}^{C,\text{nonpilot}}\right) + R_t.$$

That is, civilian compensation is the average of compensation in pilot jobs $W_{t-t\_leave}^{C,\text{nonpilot}}$ and compensation in other jobs $W_t^{C,\text{nonpilot}}$, weighted by the probability of being hired by a major airline p(hired). $W_t^{C,\text{nonpilot}}$ is the 50th percentile of pay for non-pilot veteran civilians for commissioned officers and the 70th percentile of pay for non-pilot veteran civilians for warrant officers. $W_{t-t\_leave}^{C,\text{pilot}}$ is the 70th percentile of pay for civilians in pilot jobs; t_leave is the year the individual leaves the military. We started pilot pay at the 70th percentile at age 30, regardless of when the individual leaves the Army, because pilot pay is determined almost solely by seniority within the particular airline where the pilot works (Mattock et al., 2016). We calculated percentiles of pilot and non-pilot pay using the same method presented in Appendix C and in Mattock et al. (2016).

We calculated the probability of being hired at a major airline using a logistic regression of the form

$$p(\text{hired}) = \frac{e^{a+b\times MAH}}{1 + e^{a+b\times MAH}}$$

where MAH is total MAH per year in the United States from FAPA data (see Chapter 3).

Model Estimates Including Major Airline Hiring

Table E.1 provides parameter estimates for commissioned officers when we included MAH, and Table E.2 provides the same for warrant officers. The slope for the probability of being hired by a major airline is small and not statistically significant for warrant officers and is statistically significant but substantially smaller than previous estimates for commissioned officers (see Mattock et al., 2016). For warrant officers, including MAH also makes the estimate of mean taste for the Army substantially larger.

The insignificant value for warrant officers and small value for commissioned officers suggest that major airline pay may factor into the opportunity cost for warrant officers and is only a small factor for commissioned offices. We therefore excluded the probability of being hired by a major airline from our main DRM specification.

TABLE E.1

Parameter Estimates for Commissioned Officers Including Major Airline Hiring

Parameter	Parameter Estimate	Standard Error	z-statistic	Transformed Value	Description
ln(Kappa)	5.8028	0.0628	92.4644	331.2366	Shape parameter
-ln(Mu)	0.3950	7.0501	0.0560	−1.4844	Mean taste for active service
ln(SD)	−1.1368	68.8456	−0.0165	0.3209	Standard deviation of taste for active service
-ln(Switch)	5.6661	0.0526	107.6196	−288.9006	Cost of leaving active service before end of initial ADSO
FAPA hiring intercept	1.1352	0.4936	2.2997	1.1352	Intercept for the logistic probability of MAH
FAPA hiring slope	0.19420	0.0954	2.0347	0.1942	Slope for the logistic probability of MAH
beta	0.9400	—	—	—	Assumed discount factor
Likelihood	−6,049.8430				
Observations	2,785.0000				

TABLE E.2

Parameter Estimates for Warrant Officers Including Major Airline Hiring

Parameter	Parameter Estimate	Standard Error	z-statistic	Transformed Value	Description
ln(Kappa)	5.0519	0.0422	119.8142	156.3207	Shape parameter for nest
ln(Lambda)	3.7548	0.148	25.3717	42.726	Shape parameter within nest
-ln(Mu)	−7.5003	219.7612	−0.0341	−0.0006	Mean taste for active service
ln(SD)	2.1933	0.471	4.6569	8.9646	Standard deviation of taste for active service
-ln(Switch)	4.7401	0.0383	123.8419	−114.4412	Cost of leaving active service before end of initial ADSO
FAPA hiring intercept	0.1852	0.0509	3.6374	0.1852	Intercept for the logistic probability of MAH
FAPA hiring slope	0.0055	0.0091	0.6014	0.0055	Slope for the logistic probability of MAH
beta	0.9100	—	—	—	Assumed discount factor
Likelihood	−9,458.9910				
Observations	4,117.0000				

APPENDIX F

Sensitivity Checks: Heterogeneity in Timing and Probability of Milestone Achievement

Our results in Chapter 5 on basing AvIP solely on achieving milestones depend critically on our assumptions regarding the timing and likelihood of meeting milestones. We know that our assumptions regarding a single point in time when a milestone can be gained do not truly hold, and so our calculations are an approximation only. We therefore conducted two sensitivity analyses to assess how the results would change if we varied our assumptions in the following ways:

- allowed heterogeneity in the timing of milestone achievement, but no heterogeneity in probability
- allowed heterogeneity in the probability of milestone achievement, but no heterogeneity in timing.

This appendix presents the results of these analyses.

Adding Heterogeneity in Timing and Probability of Milestone Achievement

To implement our two sensitivity checks, we assigned each simulated aviator a *type* to create a heterogenous sample. We drew types from a standard normal distribution with a mean of 0 and a standard deviation equal to 1. The type was then used to determine the timing or probability of milestone achievement in both the baseline and new policy simulations. For heterogeneity in timing, the type was rounded to the nearest integer, then that integer was subtracted from the timing of achievement for every milestone after pilot status, following Asch, Mattock, and Tong (2020).[1] For instance, individuals whose type rounds to 1 will achieve each milestone one year faster than average, whereas individuals whose type rounds to –1 will achieve each milestone one year slower than average. For heterogeneity in probability, we multiplied the probability of achievement for every milestone with achievement probability of less than 90 percent by (1 + random draw/10). Thus, for instance, individuals with a random draw of 1 will have a 10 percent higher probability of achieving milestones with an average probability of less than 90 percent (pilot-in-command, senior aviator, and master aviator for commissioned officers and track, senior aviator, and master aviator for warrant officers).

Results with Heterogeneity in Milestone Timing

The results of changing to the Army's prototype AvIP when allowing heterogeneity in milestone timing are presented in Figure F.1. The results are similar to the results when we did not allow heterogeneity (see Figure 5.1): Retention drops by 1.0 percent for commissioned officers and 1.8 percent for warrant officers.

[1] Unlike Asch, Mattock, and Tong (2020), we did not assume in this case that higher types earn more in the civilian labor market.

FIGURE F.1
The Army's Prototype AvIP with Heterogeneity in Milestone Timing

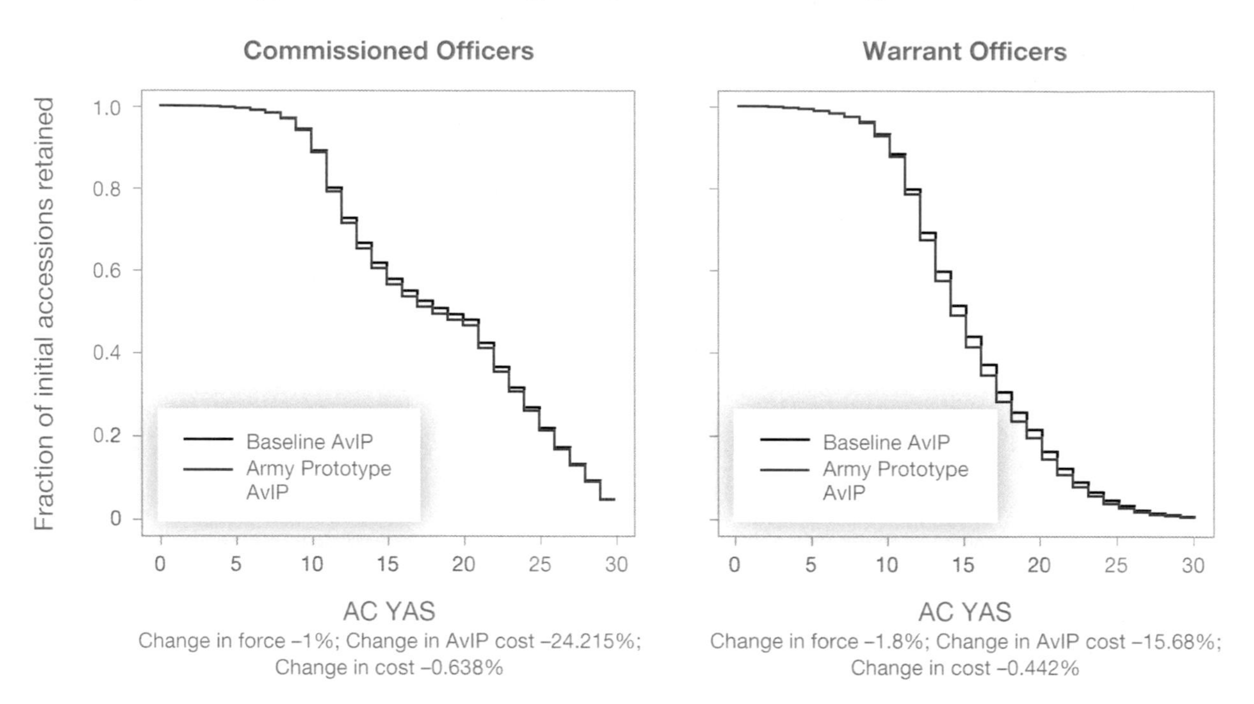

The results of allowing heterogeneity in timing of milestone achievement are presented in Table F.1. For commissioned officers, the retention-maintaining AvIP values are slightly smaller when heterogeneity in timing is allowed than when all commissioned officers gain milestones at the same time as shown in Table 5.2. We believe this means that the gain in retention for high-type officers, who gain milestones faster and whose compensation therefore increases faster than it does in the main analysis, more than offsets the loss in retention for low-type officers, who achieve milestones more slowly and whose compensation therefore increases more slowly than it does in the main analysis. However, both AvIP and overall costs rise more when there is heterogeneity in timing for commissioned officers, likely because (1) larger amounts of AvIP are now being paid to officers earlier in their career than in the main results, and (2) those more-highly-paid officers are being retained at higher rates, thus offsetting the lower values being paid for each milestone.

For warrant officers, the retention-maintaining values of AvIP are also lower in Table F.1 than in Table 5.2. Such as for commissioned officers, we believe the gain in retention for high-type warrant officers more than offsets the loss in retention for low-type warrant officers. However, the increase in costs is smaller than the main results, unlike the results for commissioned officers. This finding likely means that the values being paid for each milestone are enough lower than those in the main results to offset the fact that high-type officers are paid earlier in their career than in the main results.

Results with Heterogeneity in Milestone Achievement Probability

The results of changing to the Army's prototype AvIP when allowing heterogeneity in milestone achievement probability are presented in Figure F.2. The results are similar to the results when we did not allow heterogeneity and when we allowed heterogeneity in timing: Retention drops by 1.0 percent for commissioned officers and 1.8 percent for warrant officers.

TABLE F.1

Retention-Maintaining Values of AvIP with Heterogenous Milestone Timing

AvIP Value by Milestone	Commissioned Officers	Warrant Officers
Flight school entry	$125	$125
Pilot status	$285	$180
Pilot-in-command	$585	$415
Captain's Career Course	$1,040	$720
Senior aviator	$1,570	$1,620
Master aviator	$1,435 [a]	$1,640
% change in force size	−0.0020%	0.002%
% change in AvIP cost	14.9700%	8.060%
% change in overall cost	0.4000%	0.230%

NOTE: This table reports results of simulations of retention of commissioned officers in the Army Aviation Branch under different ways of setting AvIP, with AvIP values for each milestone reported in the columns. Baseline retention uses the AvIP values set in January 2020 (see Table 2.3). Simulations assume a ten-year initial ADSO and that AvB is not offered to warrant officers.

[a] AvIP is $0 for commissioned officers with over 25 YAS.

FIGURE F.2

The Army's Prototype AvIP with Heterogeneity in Milestone Achievement Probability

The results of allowing heterogeneity in probability of milestone achievement given timing are presented in column 4 of Table F.2 (commissioned officers) and column 3 of Table F.2 (warrant officers). For commissioned officers, the retention-maintaining AvIP values are slightly lower in Table F.2 than the retention-maintaining AvIP values without heterogeneity for pilots-in-command and master aviators in Table 5.2 (baseline), but the retention-maintaining AvIP values are higher for senior aviators and very similar (a dif-

TABLE F.2

Retention-Maintaining Values of AvIP with Heterogenous Probability of Milestone Achievement

AvIP Value by Milestone	Commissioned Officers	Warrant Officers
Flight school entry	$125	$125
Pilot status	$285	$180
Pilot-in-command	$510	$415
Captain's Career Course	$1,045	$720
Senior aviator	$1,630	$1,590
Master aviator	$1,430[a]	$1,600
% change in force size	–0.002%	0.000%
% change in AvIP cost	16.780%	6.320%
% change in overall cost	0.440%	0.200%

NOTE: This table reports results of simulations of retention of commissioned officers in the Army Aviation Branch under different ways of setting AvIP, with AvIP values for each milestone reported in the columns. Baseline retention uses the AvIP values set in January 2020 (see Table 2.3). Simulations assume a ten-year initial ADSO and that AvB is not offered to warrant officers.

[a] AvIP is $0 for commissioned officers with over 25 YAS.

ference of just $5) for the Captain's Career Course. It is possible that the rise for senior aviator is because making master aviator requires previously making senior aviator. That is, the probabilities of achieving each milestone are not independent of each other; if an officer becomes more likely to make an early milestone, then they are more likely to achieve subsequent milestones simply by virtue of meeting the prerequisites. Because the types are constant over time, the probability of meeting later milestones *conditional on meeting the prerequisites* is also higher. So, having a higher type raises the probability of making master aviator by a larger amount, in percentage point terms, than it does the probability of making pilot-in-command or senior aviator. Because senior aviators take the possibility of making master aviator in the future into account, we believe that the higher required value to maintain retention of senior aviators is a function of the drop in AvIP at the point of making master aviator.

The rise in AvIP cost is also larger with heterogeneity in probability than when there is no heterogeneity. For warrant officers, the retention-maintaining values are typically lower when heterogeneity in probability is allowed than when all warrant officers gain milestones with the same probability, and the rise in AvIP costs is lower. It seems likely that the drop is because the higher retention of individuals who are more likely to make the milestones offsets the lower retention of individuals who are less likely to make the milestones.

Policy Implications of Sensitivity Checks

What did we learn from the sensitivity checks presented in this appendix? The main implication that we took from these analyses is that our results are sensitive to our assumptions on the timing and probability of milestone achievement. The retention-maintaining values of monthly AvIP do not shift by large amounts—in most cases, the differences between values found in this appendix for milestones prior to master aviator do not differ by more than $50, and for master aviator (a milestone achieved by only a small minority of personnel), the range of differences in values is $120. While these are relatively small differences, they are large enough for us to note that our values for retention-maintaining monthly AvIP amounts depend on our modeling assumptions and should therefore be taken as an approximation only.

Abbreviations

AC	active component
ACS	American Community Survey
ADSO	active-duty service obligation
AFS	active federal service
ALARACT	All Army Activities
AR	Army Regulation
ARNG	Army National Guard
AvIP	aviation incentive pay
AvB	Aviation Bonus
BRS	blended retirement system
CAB	combat aviation brigade
COVID-19	coronavirus disease 2019
CPI-U	Consumer Price Index for All Urban Consumers
CW	chief warrant officer
DMDC	Defense Manpower Data Center
DoD	U.S. Department of Defense
DoDI	Department of Defense Instruction
DRM	dynamic retention model
FAA	Federal Aviation Administration
FAPA	Future and Active Pilot Advisors
FY	fiscal year
IERW	Initial Entry Rotary Wing
MAH	major airline hiring
MILPER	Military Personnel (Message)
MOS	military occupational specialty
MTOE	Modified Table of Organization and Equipment
MTFE	maintenance test flight examiner
MTP	maintenance test pilot
OSD	Office of the Secretary of Defense
PME	professional military education

QRMC	Quadrennial Review of Military Compensation
RA	Regular Army
RMC	regular military compensation
RTP	rotary transition program
S&I	special and incentive
SQI	special qualifications identifier
TAPDB	Total Army Personnel Database
TDA	Table of Distribution and Allowances
TSP	Thrift Savings Plan
UND	University of North Dakota
UPS	United Parcel Service
USAR	U.S. Army Reserve
VAC	Veteran to Aviation Charity
WO1	warrant officer 1
YAS	year(s) of aviation service
YOS	year(s) of service

Glossary

These definitions are drawn from AR 600-105 (2020).

Aeronautical rating: One of several qualifications awarded to officers. A rating certifies successful completion of prescribed aviation-related training or equivalent experience. Current Army aeronautical ratings are Army aviator, senior Army aviator, master Army aviator, flight surgeon, senior flight surgeon, and master flight surgeon.

Army aviator: Officers who have been awarded an Army aeronautical rating. This rating does not include flight surgeons.

Aviation gates: The 12th and 18th year computed from the aviation service entry date in an aviator's aviation career used to determine further eligibility for AvIP. Gates do not apply to flight surgeons that follow conditional AvIP rules.

Career AvIP: Additional pay intended to attract and retain officers in an aviation career field on a career basis. It is authorized to aviators in a designated career field that remain in aviation service on a career basis. AvIP is not directly linked to monthly flying requirements but instead is linked to meeting the career flying duty requirements in paragraph 3-1 of AR 600-105, 2020.

Conditional career AvIP: Pay authorized to flight surgeons while practicing aviation medicine in an authorized operational position and aviators who are assigned to operational flying duty and are performing the minimum flight requirements in a military aircraft but are not otherwise eligible for career AvIP according to paragraph 3-1b of AR 600-105, 2020.

Officer: The term officer is used in this report to indicate WO1 through general officer aviators and flight surgeons. When the need exists, they are separated but are otherwise interchangeable.

Pilot status codes (quoted from AR 600-105, 2020, p. 30): Codes used to identify aviators on the Officer Record Brief, as follows:

- *Pilot Status Code 1*: Qualified for aviation service
- *Pilot Status Code 2*: Medically disqualified
- *Pilot Status Code 3*: Non-medically disqualified (by flying evaluation board or voluntary or involuntary disqualification)
- *Pilot Status Code 4*: Not in aviation service
- *Pilot Status Code 5*: Conditional AvIP.

References

Airline Pilot Central, homepage, undated. As of February 8, 2023:
https://www.airlinepilotcentral.com/

ALARACT—*See* All Army Activities.

All Army Activities 071/2018, *Army Aviation Bonus (AvB) Program*, Department of the Army, August 2018.

All Army Activities 075/2019, *U.S. Army Conventional Force Aviation Bonus (AvB) Program*, Department of the Army, October 2019.

All Army Activities 085/2017, *Army Aviation Bonus (AvB) Program*, Department of the Army, September 2017.

American Airlines, "American Airlines Reports Third-Quarter 2021 Financial Results," October 21, 2021.

Army Regulation 95-1, *Aviation Flight Regulations*, Headquarters, Department of the Army, March 22, 2018.

Army Regulation 600-105, *Aviation Service of Rated Army Officers*, Headquarters, Department of the Army, June 5, 2020.

AR—*See* Army Regulation.

Asch, Beth J., James Hosek, Michael Mattock, and Christina Panis, *Assessing Compensation Reform: Research in Support of the 10th Quadrennial Review of Military Compensation*, RAND Corporation, MG-764-OSD, 2008. As of July 13, 2022:
https://www.rand.org/pubs/monographs/MG764.html

Asch, Beth J., Michael G. Mattock, and James Hosek, *The Blended Retirement System: Retention Effects and Continuation Pay Cost Estimates for the Armed Services*, RAND Corporation, RR-1887-OSD/USCG, 2017. As of August 25, 2022:
https://www.rand.org/pubs/research_reports/RR1887.html

Asch, Beth J., Michael G. Mattock, James Hosek, and Shanthi Nataraj, *Assessing Retention and Special and Incentive Pays for Army and Navy Commissioned Officers in the Special Operations Forces*, RAND Corporation, RR-1796-OSD, 2019. As of September 15, 2022:
https://www.rand.org/pubs/research_reports/RR1796.html

Asch, Beth J., Michael G. Mattock, James Hosek, and Patricia K. Tong, *Capping Retired Pay for Senior Field Grade Officers: Force Management, Retention, and Cost Effects*, RAND Corporation, RR-2251-OSD, 2018. As of July 13, 2022:
https://www.rand.org/pubs/research_reports/RR2251.html

Asch, Beth J., Michael G. Mattock, and Patricia K. Tong, *Analysis of a Time-in-Grade Pay Table for Military Personnel and Policy Alternatives*, RAND Corporation, RR-A369-1, 2020. As of July 19, 2022:
https://www.rand.org/pubs/research_reports/RRA369-1.html

Asch, Beth J., Michael G. Mattock, Patricia K. Tong, and Jason M. Ward, *Increasing Efficiency and Incentives for Performance in the Army's Selective Reenlistment Bonus (SRB) Program*, RAND Corporation, RR-A803-1, 2021. As of August 25, 2022:
https://www.rand.org/pubs/research_reports/RRA803-1.html

Asch, Beth J., and John T. Warner, A Policy Analysis of Alternative Military Retirement Systems, RAND Corporation, MR-465-OSD, 1994. As of August 28, 2019:
https://www.rand.org/pubs/monograph_reports/MR465.html

Asch, Beth J., and John T. Warner, "A Theory of Compensation and Personnel Policy in Hierarchical Organizations with Application to the United States Military," *Journal of Labor Economics*, Vol. 19, No. 3, July 2001.

ATP, "Commercial Airline Pilot Salary," webpage, undated. As of September 15, 2022:
https://atpflightschool.com/become-a-pilot/airline-career/commercial-pilot-salary.html

Aviation Technician Education Council, "Boeing Releases 2019-2038 Technician Outlook," *ATEC News*, July 1, 2019.

AVweb, "Great Time to Be A Pilot: Boeing," April 12, 2019.

Baldanza, Ben, "Yes, a Pilot Shortage Is Real—What Airlines Can Do While Raising Pay," *Forbes*, September 1, 2022.

Boeing, "Boeing Pilot and Technician Outlook 2020–2039," fact sheet, 2020.

Boeing, "Boeing Pilot and Technician Outlook 2021–2040," fact sheet, 2021.

Boeing, "Boeing Pilot and Technician Outlook 2022–2041," fact sheet, 2022.

Code of Federal Regulations, Title 14, Aeronautic and Space; Chapter I, Federal Aviation Administration, Department of Transportation; Subchapter D, Airmen; Section 61.160, Aeronautical Experience—Airplane Category Restricted Privileges.

Crookston, Jaidyn, "Is It Easy to Get a Job as a Helicopter Pilot?" Southern Utah University, May 18, 2021.

Defense Advisory Committee on Military Compensation, *The Military Compensation System: Completing the Transition to an All-Volunteer Force, Report of the Defense Advisory Committee on Military Compensation*, April 28, 2006.

Delta Air Lines, "Delta Air Lines Announces September Quarter 2021 Profit," October 13, 2021.

Department of Defense Instruction 7000.14-R, Volume 7A, Chapter 22, *Aerial Flights*, U.S. Department of Defense, February 2001.

Department of Defense Instruction 7000.14-R, Volume 7A, Chapter 22, *Aviation Incentive Pays*, U.S. Department of Defense, August 2022.

Department of Defense Office of the Actuary, "Valuation of the Military Retirement System" U.S. Department of Defense, September 30, 2020.

Department of the Army Pamphlet 600-3, *Officer Professional Development and Career Management*, Department of the Army, April 3, 2022.

DoD—*See* U.S. Department of Defense.

DoDI—*See* Department of Defense Instruction.

Envoy, "Rotor Transition Program," webpage, undated. As of September 19, 2022:
https://www.envoyair.com/pilots/rotor-transition-program/

Everyday Aviation, "Helicopter Pilot Salary," webpage, undated. As of September 19, 2022:
https://everydayaviation.com/helicopter-pilot-salary/

FAA—*See* Federal Aviation Administration.

FAPA—*See* Future and Active Pilot Advisors.

Federal Aviation Administration, *FAA Aerospace Forecast Fiscal Years 2021–2041*, 2021.

FedEx, "FedEx Corp. Reports Strong Third Quarter Results," press release, October 2021.

FRED, "Consumer Price Index for All Urban Consumers: All Items in U.S. City Average," dataset, undated-a. As of April 13, 2023:
https://fred.stlouisfed.org/series/CPIAUCSL

FRED, "Unemployment Rate—Bachelor's Degree and Higher, 25 Yrs. And Over," dataset, undated-b. As of April 13, 2023:
https://fred.stlouisfed.org/series/LNS14027662

Future and Active Pilot Advisors, "Major Airline Pilot Hiring by Year (2000-Present),"dataset, undated. As of September 15, 2022:
https://fapa.aero/hiringhistory.asp

Gordon, Lisa, "Study Confirms Looming Helicopter Pilot and Maintainer Shortage," *Vertical*, March 1, 2018.

Gotz, Glenn A., and John J. McCall, *A Dynamic Retention Model for Air Force Officers: Theory and Estimates*, RAND Corporation, R-3028-AF, 1984. As of September 26, 2022:
https://www.rand.org/pubs/reports/R3028.html

Greenberg, Peter, "Why Is There a Pilot Shortage? It Wasn't Just the COVID-19 Pandemic," CBS News, July 21, 2022.

Holistic Aviation Assessment Task Force, "Regaining Decisive Action Readiness," May 1, 2016.

Hosek, James, Shanthi Nataraj, Michael G. Mattock, and Beth J. Asch, *The Role of Special and Incentive Pays in Retaining Military Mental Health Care Providers*, RAND Corporation, RR-1425-OSD, 2017. As of September 15, 2022:
https://www.rand.org/pubs/research_reports/RR1425.html

Infinity Flight Group, "Overview," webpage, undated. As of September 19, 2022:
https://infinityflightgroup.com/helicopter-transition/

Joseph, Leslie, "Airline Employees' Dilemma: Take Severance or Gamble on Oct. 1 Layoffs," CNBC, June 9, 2020.

Joseph, Leslie, "A Severe Pilot Shortage in the U.S. Leaves Airlines Scrambling for Solutions," CNBC, May 15, 2022a.

Joseph, Leslie, "American Airlines Regional Carriers Hike Pilot Pay More Than 50% as Shortage Persists," CNBC, June 13, 2022b.

Lankford, Kimberly, "A Fighter Pilot's New Career Takes Off," Kiplinger, February 23, 2017.

Mattock, Michael, and Jeremy Arkes, *The Dynamic Retention Model for Air Force Officers: New Estimates and Policy Simulations of the Aviator Continuation Pay Program*, RAND Corporation, TR-470-AF, 2007. As of July 13, 2022:
https://www.rand.org/pubs/technical_reports/TR470.html

Mattock, Michael G., Beth J. Asch, James Hosek, and Michael Boito, *The Relative Cost-Effectiveness of Retaining Versus Accessing Air Force Pilots*, RAND Corporation, RR-2415-AF, 2019. As of November 1, 2022:
https://www.rand.org/pubs/research_reports/RR2415.html

Mattock, Michael G., James Hosek, and Beth J. Asch, *Reserve Participation and Cost Under a New Approach to Reserve Compensation*, RAND Corporation, MG-1153-OSD, 2012. As of September 15, 2022:
https://www.rand.org/pubs/monographs/MG1153.html

Mattock, Michael G., James Hosek, Beth J. Asch, and Rita Karam, *Retaining U.S. Air Force Pilots When the Civilian Demand for Pilots Is Growing*, RAND Corporation, RR-1455-AF, 2016. As of July 19, 2022:
https://www.rand.org/pubs/research_reports/RR1455.html

McGee, Michael, *Air Transport Pilot Supply and Demand: Current State and Effects of Recent Legislation*, dissertation, Pardee RAND Graduate School, RAND Corporation, RGSD-351, 2015. As of September 15, 2022:
https://www.rand.org/pubs/rgs_dissertations/RGSD351.html

Military Personnel Message 06-318, *FY07 Aviation Continuation Pay (ACP) Program (Special Operations Aviators (SOAR))*, Department of the Army, November 3, 2006.

Military Personnel Message 06-319, *FY07 Aviation Continuation Pay (ACP) Program (Maintenance Test Pilots (MTP) and Maintenance Test Flight Examiners (MTFE))*, Department of the Army, November 3, 2006.

Military Personnel Message 07-305, *FY08 Aviation Continuation Pay (ACP) Program (Maintenance Test Pilots (MTP) and Maintenance Test Flight Examiners (MTFE))*, Department of the Army, November 8, 2007.

Military Personnel Message 07-306, *FY08 Aviation Continuation Pay (ACP) Program (Special Operations Aviators (SOAR))*, Department of the Army, November 8, 2007.

Military Personnel Message 07-307, *FY08 Aviation Continuation Pay (ACP) Program (Tactical Operations Warrant Officers (TACOPS- SQI I))*, Department of the Army, November 8, 2007.

Military Personnel Message 08-181, *FY08 Aviation Continuation Pay (ACP) Program for 152F/H and 154C/154F Aviators with Targeted Basic Active Service Date (BASD)—19 to 24 Years—and Special Qualification Identifiers (SQI-C, F, H, G, L) (REVISED)*, Department of the Army, July 2, 2008.

Military Personnel Message 08-256, *FY09 Aviation Continuation Pay (ACP) Program (Maintenance Test Pilots (MTP) and Maintenance Test Flight Examiners (MTFE))*, Department of the Army, October 9, 2008.

Military Personnel Message 08-257, *FY09 Aviation Continuation Pay (ACP) Program (Special Operations Aviators (SOAR))*, Department of the Army, October 9, 2008.

Military Personnel Message 08-258, *FY09 Aviation Continuation Pay (ACP) Program (Tactical Operations Warrant Officers (TACOPS-SQI I))*, Department of the Army, October 9, 2008.

Military Personnel Message 09-246, *FY10 Aviation Continuation Pay (ACP) Program (Special Operations Aviation Regiment (SOAR))*, Department of the Army, October 22, 2009.

Military Personnel Message 11-053, *FY11 Aviation Continuation Pay (ACP) Program (Special Operations Aviation Regiment (SOAR))*, Department of the Army, February 23, 2011.

Military Personnel Message 11-054, *FY11 Aviation Continuation Pay (ACP) Program for 152F/H Aviators with Targeted Basic Active Service Date (BASD)—19 to 24 Years—and Special Qualification Identifiers (SQI-C, F, H)*, Department of the Army, February 22, 2011.

MILPER—*See* Military Personnel Message.

Office of the Under Secretary of Defense for Personnel and Readiness, "Selected Military Compensation Tables," datasets, January 2000–January 2019.

Office of the Under Secretary of Defense for Personnel and Readiness, *Report to Congressional Armed Services Committees on Initiatives for Mitigating Military Pilot Shortfalls*, January 16, 2019.

OUSD P&R—*See* Office of the Under Secretary of Defense for Personnel and Readiness.

Philpott, Tom, "Final Opt-In Rates for Blended Retirement Yield More Surprises," *Stars and Stripes*, January 17, 2019.

Rempfer, Kyle, "New Army Aviators Will Incur 10-Year Service Obligations, Up from Six, Starting in October," *Army Times*, August 21, 2020.

Rotary to Airline Group (RTAG), "RTAG Leadership," webpage, undated. As of September 19, 2022:
https://www.rtag.org/about/leadership

Ruggles, Steven, Catherine A. Fitch, Ronald Goeken, J. David Hacker, Matt A. Nelson, Evan Roberts, Megan Schouweiler, and Matthew Sobek, "IPUMS Ancestry Full Count Data: Version 3.0," dataset, IPUMS, 2021.

Ruggles, Steven, Sarah Flood, Ronald Goeken, Megan Schouweiler and Matthew Sobek, "IPUMS USA: Version 12.0," dataset, IPUMS, 2022.

Salary.com, "Chief Helicopter Pilot Salary in the United States," webpage, August 29, 2022a. As of September 19, 2022:
https://www.salary.com/research/salary/benchmark/chief-helicopter-pilot-salary

Salary.com, "Helicopter Pilot Salary in the United States," webpage, August 29, 2022b. As of September 19, 2022:
https://www.salary.com/research/salary/benchmark/helicopter-pilot-salary

Schaper, David, "Proposals Would Ease Standards, Raise Retirement Age to Address Pilot Shortage," NPR, August 10, 2022.

Southwest Airlines, "Southwest Airlines to Discuss Third Quarter 2021 Financial Results on October 21, 2021," press release, October 2021.

Talent.com, "Helicopter Pilot Average Salary in the USA 2022," webpage, undated. As of September 19, 2022:
https://www.talent.com/salary?job=helicopter+pilot

Terry, Tara L., Jeremy M. Eckhause, Michael McGee, James H. Bigelow, and Paul Emslie, *Projecting Air Force Rated Officer Inventory Across the Total Force: Total Force Blue Line Model for Rated Officer Management*, RAND Corporation, RR-2796, 2019. As of September 15, 2022:
https://www.rand.org/pubs/research_reports/RR2796.html

Thrift Savings Plan Bulletin 20-7, *Implementation of 5% Automatic Enrollment Percentage for Thrift Savings Plan Participants*, September 23, 2020.

TSP—*See* Thrift Savings Plan.

Under Secretary of Defense for Personnel and Readiness, *Military Compensation Background Papers Compensation Elements and Related Manpower Cost Items: Their Purposes and Legislative Backgrounds*, 8th ed., U.S. Department of Defense, July 2018.

United Airlines, "United Airlines Third-Quarter Results, Remains on Track to Meet 2022 Targets, Poised to Capitalize on International Reopening," press release, October 2021.

U.S. Army, "Warrant Officers," webpage, undated. As of September 19, 2022:
https://m.goarmy.com/careers-and-jobs/current-and-prior-service/advance-your-career/warrant-officer.m.html

U.S. Army Human Resources Command, "Aviation Bonus Program—FY2022 Program," April 12, 2022.

U.S. Bureau of Labor Statistics, "American Community Survey," undated.

U.S. Bureau of Labor Statistics, "Occupational Employment and Wages, May 2021," webpage, March 31, 2022. As of September 19, 2022:
https://www.bls.gov/oes/current/oes532012.htm#ind